Painting for
Collision Repair

Michael Crandell
Carl Sandburg College

Delmar Publishers

an International Thomson Publishing company I(T)P®

Albany • Bonn • Boston • Cincinnati • Detroit • London • Madrid
Melbourne • Mexico City • New York • Pacific Grove • Paris • San Francisco
Singapore • Tokyo • Toronto • Washington

NOTICE TO THE READER

Cover Design: Paul Roseneck

Delmar Staff
Publisher: Alar Elken
Acquisitions Editor: Vernon Anthony
Editorial Assistant: Betsy Hough
Fast Cycle Production Editor: Dianne Jensis
Marketing Manager: Mona Caron

COPYRIGHT © 1999
By Delmar Publishers

an International Thomson Publishing company I**T**P®

The ITP logo is a trademark under license.
Printed in the United States of America

Online Services

Delmar Online
To access a wide variety of Delmar products and services on the World Wide Web, point your browser to:
   http://www.delmar.com
   or email: info@delmar.com

A service of I**T**P

For more information, contact:

Delmar Publishers
3 Columbia Circle, Box 15015
Albany, New York 12212-5015

International Thomson Publishing Europe
Berkshire House
168-173 High Holborn
London, WC1V7AA
United Kingdom

Nelson ITP, Australia
102 Dodds Street
South Melbourne,
Victoria, 3205 Australia

Nelson Canada
1120 Birchmont Road
Scarborough, Ontario
M1K 5G4, Canada

International Thomson Publishing France
Tour Maine-Montparnasse
33 Avenue du Maine
75755 Paris Cedex 15, France

International Thomson Editores
Seneca 53
Colonia Polanco
11560 Mexico D. F. Mexico

International Thomson Publishing GmbH
Königswinterer Strasse 418
53227 Bonn
Germany

International Thomson Publishing Asia
60 Albert Street
#15-01 Albert Complex
Singapore 189969

International Thomson Publishing Japan
Hirakawa-cho Kyowa Building, 3F
2-2-1 Hirakawa-cho, Chiyoda-ku,
Tokyo 102, Japan

ITE Spain/Paraninfo
Calle Magallanes, 25
28015-Madrid, Spain

1 2 3 4 5 6 7 8 9 10 XXX 03 02 01 00 99 98

Library of Congress Cataloging-in-Publication Data

Crandell, Michael.
     Painting for collision repair / Michael Crandell.
          p.   cm.
     Includes index.
     ISBN 0–7668–0905–6
     1. Automobiles—Painting. 2. Automobiles—Collision damage.
     I. Title.
     TL255.2.C73   1998
     629.2'6'0288—dc21
                                        98–34630
                                          CIP

Contents

Preface

This book is a compilation of my experiences in autobody repair. In the twenty-plus years I have painted cars, I have seen many changes in the industry. Perhaps the greatest change is the acceptance of the fact that paint vapors are dangerous. In the 1970s, the painters in the shop I worked at, myself included, did not wear respirators, ever. Some of the painters reported discomfort 8 to 10 hours after spraying paint that contained flex agent. We did not know it at the time, but these painters were experiencing isocyanate sensitization. Today, the dangers of paint spraying are known. The most important advice I can give to a beginning painter is to protect your health. I do not know any painter from the primitive days who lived to the age of sixty.

As you read this book, understand that there are many opinions on how to repair a vehicle. The repaired vehicle should have the same appearance and durability as an undamaged vehicle. Every step in the repair should be directed toward this goal. This book explains the series of steps I use to repair a vehicle. The reader should consult the paint manufacturer's refinish manual for specific instructions.

Vehicle refinishing can be a rewarding career. The skill to spray a beautiful topcoat may take years to acquire. Read this book and practice. Mistakes like runs, excessive orange peel, and mottling are common to beginning painters. Do not be overly concerned about mistakes; there is a solution for every refinishing mistake.

Acknowledgments

The author would like to thank all of the employees at the following companies for their help in making this book possible.

• Dave's Autobody	Galesburg, Illinois
• Uftring Chevrolet	Washington, Illinois
• Pete's Autobody	Galesburg, Illinois
• Automotive Finishes	Peoria, Illinois
• Big A Auto Parts	Galesburg, Illinois

I also want to acknowledge my autobody repair students, especially Neil Rinkack, for helping with many of the photographs.

Of course, thanks to my wife, Janine, for typing the manuscript and doing the initial editing.

The following companies contributed photographs and diagrams.

- Binks Mfg. Co.
- DuPont Automotive Refinish Products
- DuPont Company
- ITW DeVilbiss Automotive Refinishing Products
- Mitchell International
- PPG
- Sharp Manufacturing Company
- Toyota Motor Corporation

Chapter 1
Safety Procedures

Objectives

After reading this chapter, you will be able to:
- Describe the various precautions for personal safety that must be followed in a collision repair facility.
- Outline several health and safety measures necessary when working with hazardous vapors and solvents.
- List the types of hazardous wastes.
- Describe safety practices to avoid fire and explosion.
- List the classifications of fires.

Key Term List

OSHA
MSDS
isocyanate
vapor
solvent
CO
EPA
hazardous waste
VOC
ASE

1.1 SHOP SAFETY PRACTICES

The most important considerations in any refinishing shop should be accident prevention and safety. Carelessness and the lack of safety habits cause accidents. Accidents have a far-reaching effect, not only on the victim, but also on the victim's family and society in general. More importantly, accidents can cause serious injury, temporary or permanent, or even death. Therefore, it is the obligation of all shop employees and the employer to foster and develop a safety program to protect the health and welfare of those involved.

In this book, the text contains special notations labeled **SHOP TALK,** *CAUTION,* and *WARNING.* Each one is there for a specific purpose. **SHOP TALK** gives added information that will help the technician to complete a particular procedure or make a task easier. *CAUTION* is given to prevent the technician from making an error that could damage the vehicle. *WARNING* reminds the technician to be especially careful of those areas where carelessness can cause personal injury. The following text contains some general *WARNINGS* that should be followed when working in a collision repair and paint facility.

MANUFACTURERS' WARNINGS AND GOVERNMENT REGULATIONS

Most of the products used in a collision repair facility carry warning and caution information that must be read and understood by all users before use. Likewise, all federal (including Occupational Safety and

Health Administration [**OSHA**], Mine Safety and Health Administration [MSHA], and National Institute for Occupational Safety and Health [NIOSH]), state, and local safety regulations should not only be *fully* understood, but strictly observed.

Most paint products are hazardous materials. Every employee in the shop is protected by right-to-know laws. These laws started with OSHA's Hazard Communication Standard published in 1983. This document was originally intended for chemical companies and manufacturers that require employees to handle potentially hazardous materials in the workplace. Since then, the majority of states have enacted their own right-to-know laws, and the federal courts have decided that these regulations should apply to all companies, including the auto refinishing profession.

The general intent of the laws is for employers to provide their employees with a safe working place as it relates to hazardous materials. Specifically, there are three areas of employer responsibility:

1. **Training/educating employees.** All employees must be trained about their rights under the legislation, the nature of the hazardous chemicals in their workplace, the labeling of chemicals, and the information about each chemical posted on Material Safety Data Sheets (**MSDS**) (Figure 1–1). These sheets detail product composition and precautionary information for all products that can present a health or safety hazard. These sheets or forms are generally prepared by the material supplier. Employees must be familiarized with the general uses, characteristics, protective equipment, accident or spill procedures, and so on associated with major groups of chemicals.
2. **Labeling/information about potentially hazardous chemicals.** All hazardous materials must be properly labeled, indicating what health, fire, or reactivity hazard they pose and what protective equipment is necessary when handling each chemical. The manufacturer of the hazardous materials must provide all warnings and precautionary information, which must be read and understood by the user before application. Attention to all label precautions is essential to the proper use of the material and the prevention of hazardous conditions.
3. **Record keeping.** Shops must maintain documentation on the hazardous chemicals in the workplace, proof of training programs, records of accidents and/or spill incidents, satisfaction of employee requests for specific chemical information via the MSDSs, and a general right-to-know compliance procedure manual utilized within the shop.

PERSONAL SAFETY AND HEALTH PROTECTION

The following are very important personal safety rules that must be heeded.

LUNG PROTECTION

Abrasive dust, **vapors** from caustic solutions and solvents, spray mist from undercoats and finishes—all present dangers to the air passages and lungs, especially for workers who are among them day in, day out.

Solvent vapors are present when mixing or spraying paint. Inhaling these vapors is not only harmful to the lungs, it places the chemicals in the vapors directly into the bloodstream. These harmful chemicals include isocyanates, xylene, and toluene. Once in the bloodstream these chemicals may travel throughout the body, possibly reaching the brain, kidneys, heart, or liver. Depending on the type of chemical, the symptoms of exposure could be memory loss, dizziness, or vomiting. Long-term exposure to these chemicals could lead to nervous system dysfunction, impaired pulmonary function, or cancer. An individual exposed to isocyanates may have an acute respiratory allergic reaction (also known as sensitized to isocyanates). Once an individual is sensitized to isocyanates and is exposed again, he or she may have a severe reaction to even low levels of isocyanates. A sensitized individual should not work in an environment where isocyanate exposure is possible. Respirators are needed in refinishing shops even though adequate ventilation is provided for the work areas. There are three primary types of respirators available to protect refinishing technicians: the supplied air respirator, cartridge filter respirator, and dust or particle respirator. Be sure to check the MSDS sheet to see what type of respirator must be used for a particular paint product.

Air-Supplied or Hood Respirators

A NIOSH approved air-supplied respirator provides protection from sensitization and other dangers of

MSDS NO. 19-1
REFINISH SALES
JANUARY 1, 1998

MATERIAL SAFETY DATA SHEET

CHROMABASE® CLEAR, ACTIVATOR, REDUCERS

Section I - Manufacturer

Manufacturer:
DuPont Co.
Automotive
Wilmington, Delaware 19898
Telephone:
Product information (800)441-7515
Medical emergency (800) 441-3637
Transportation emergency (800) 424-9300 (CHEMTREC)
Product: Chromabase® Clear, Activator, Reducers(7500S, 7565S, 7575S, 7585S, 7595S, 7600S, 7655S, 7675S, 7695S, 7800S, 7875S, 7895S).
OSHA Hazard Class: Flammable liquid
DOT Shipping Name: Paint, UN1263; Paint Related Material, UN1263
Hazardous Materials Information: See Section X.

Section II - Hazardous Ingredients
(See Section X)

Ingredients CAS No.		Vapor Pressure (20°C. mm Hg)	Exposure Limits
Acetic acid ester of C9-C11 oxo-alcohol			
	108419-34-7	0.1@ 21°C	50 ppm-S
			None-A,O
Acrylic polymer -A			
	69215-54-9	None	None-A,O
Acrylic polymer -B			
	Not Available	None	None-A,O
Acrylic polymer -C			
	833-70-5	None	None-A,O
Aliphatic polyisocyanate resin-A			
	28182-81-2	None	0.5 mg/m³-S
			1 mg/m³-S 15 min(STEL)
			None-A,O
Aliphatic polymeric isocyanate-B			
	3779-63-3	Unknown	0.5 mg/m³-S 8 hr TWA
			1 mg/m³-S 15 min(STEL)
			None-A,O
Aromatic hydrocarbon			
	64742-95-6	10.0 @ 25°C	None-A,O
Butyl acetate			
	123-86-4	8.0	150 ppm-A,O
			200 ppm-A 15 min(STEL)
Cumene	98-82-8	3.7	50 ppm-A,O Skin
Ethyl acetate			
	141-78-6	76.0	400 ppm-A,O
Ethyl 3-ethoxy propionate			
	763-69-9	Unknown	None-A,O
Ethylbenzene			
	100-41-4	7.0	100 ppm-A,O
			125 ppm-A 15 min(STEL)
Ethylene glycol monobutyl ether acetate			
	112-07-2	0.3	20 ppm-D Skin
			None-A,O
Hexyl acetate isomers			
	88230-35-7	0.7	50 ppm-A Hexyl Acet
			None-O
Methyl ethyl ketone			
	78-93-3	71.0	200 ppm-A,O
			300 ppm-A 15 min(STEL)
			200 ppm-D 8&12 hr TWA
			300 ppm-D 15 min TWA
Methyl isobutyl ketone			
	108-10-1	15.0	50 ppm-A
			100 ppm-O
			75 ppm-A 15 min(STEL)

Polyester resin			
	65086-73-9	None	None-A,O
Polyester resin			
	Not Available	None	None-A,O
Propylene glycol monomethyl ether acetate			
	108-65-6	3.7	None-A,O
			10 ppm-D
Toluene	108-88-3	36.7	50 ppm-A
			200 ppm-O
			300 ppm-O Ceiling
			500 ppm-O 10 min MAX
			50 ppm-D 8&12 hr TWA
Xylene	1330-20-7	7.0 @ 25°C	100 ppm-A,O
			150 ppm-A (STEL)
			100 ppm-D 8&12 hr
			150 ppm-D 15 min TWA
1,2,4-Trimethyl Benzene			
	95-63-6	7.0 @ 44.4°C	25 ppm-A,O
1,6-hexamethylene diisocyanate			
	822-06-0	Unknown	5 ppb-A
			None-O

A = ACGIH TLV; O= OSHA; D = DuPont internal limit; S= Supplier Furnished limit; STEL = Short Term Exposure Limit; C= Ceiling.

Section III - Physical Data

Evaporation rate: Less than ether
Vapor Density: Heavier than air
Solubility in water: Miscible
Percent volatile by volume: 34.3%- 72.8%
Percent volatile by weight: 28.7%- 67.1%
Boiling range: 76°C- 249°C/ 169°F- 480°F
Gallon weight: 7.75 - 9.02 lb/gallon

Section IV - Fire and Explosion Data

Flash point (closed cup): See Section X for exact values.
Flammable limits: 0.8%- 13.1%
Extinguishing media: Universal aqueous film-forming foam, carbon dioxide, dry chemical.
Special fire fighting procedures: Full protective equipment, including self-contained breathing apparatus, is recommended. Water from fog nozzles may be used to cool closed containers to prevent pressure build up.
Unusual fire & explosion hazards: When heated above the flash point, emits flammable vapors which, when mixed with air, can burn or be explosive. Fine mists or sprays may be flammable at temperatures below the flash point.

Section V - Health Hazard Data

General Effects:
Ingestion: Gastrointestinal distress. In the unlikely event of ingestion, call a physician immediately and have the names of ingredients available. **DO NOT INDUCE VOMITING.**
Inhalation: May cause nose and throat irritation. Repeated and prolonged overexposure to solvents may lead to permanent brain and nervous system damage. Eye watering, headaches, nausea, dizziness and loss of coordination are signs that solvent levels are too high. Exposure to isocyanates may cause respiratory sensitization. This effect may be permanent. This effect may be delayed for several hours after exposure. Repeated overexposure to isocyanates may cause a decrease in lung function which may be permanent. Individuals with or breathing problems or prior reaction to isocyanates must not be exposed to vapors or spray mist of this product. If affected by inhalation of vapor or spray mist, remove to fresh air. If breathing difficulty persists, or occurs later, consult a physician.
Skin or eye contact: May cause irritation or burning of the eyes. Repeated or prolonged liquid contact may cause skin irritation with discomfort and dermatitis. In case of eye contact, immediately flush with plenty of water for at least 15 minutes; call a physician. In case of skin contact, wash with soap and water. If irritation occurs, contact a physician.

continued

FIGURE 1–1 MSDS sheet for Du Pont clear. (Courtesy of Du Pont Co.)

Specific Effects:
Aliphatic polyisocyanate resin Repeated exposure may cause allergic skin rash, itching, swelling. May cause eye irritation with discomfort, tearing, or blurred vision. Repeated overexposure to isocyanates may cause lung injury, including a decrease in lung function, which may be permanent. Overexposure may cause asthma-like reactions with shortness of breath, wheezing, cough, which may be permanent; or permanent lung sensitization. This effect may be delayed for several hours after exposure. Individuals with preexisting lung disease, asthma or breathing difficulties may have increased susceptibility to the toxicity of excessive exposures. **Aliphatic polymeric isocyanate** Repeated exposure may cause allergic skin rash, itching, swelling. Repeated overexposure to isocyanates may cause lung injury, including a decrease in lung function, which may be permanent. May cause eye irritation with discomfort, tearing, or blurred vision. Overexposure may cause asthma-like reactions with shortness of breath, wheezing, cough, which may be permanent; or permanent lung sensitization. This effect may be delayed for several hours after exposure. Individuals with preexisting lung disease, asthma or breathing difficulties may have increased susceptibility to the toxicity of excessive exposures. **Aromatic hydrocarbon** Laboratory studies with rats have shown that petroleum distillates can cause kidney damage and kidney or liver tumors. These effects were not seen in similar studies with guinea pigs, dogs, or monkeys. Several studies evaluating petroleum workers have not shown a significant increase of kidney damage or an increase in kidney or liver tumors. **Butyl acetate** May cause abnormal liver function. Tests for embryotoxic activity in animals has been inconclusive. Has been toxic to the fetus in laboratory animals at doses that are toxic to the mother. **Ethyl acetate** Prolonged and repeated high exposures of laboratory animals resulted in secondary anemia with an increase in white blood cells; fatty degeneration, cloudy swelling and an excess of blood in various organs. **Ethyl 3-ethoxy propionate** Has been toxic to the fetus in laboratory animals at doses that are toxic to the mother. **Ethylbenzene** Recurrent overexposure may result in liver and kidney injury. Studies in laboratory animals have shown reproductive, embryotoxic and developmental effects. Has shown mutagenic activity in laboratory cell culture tests. Tests in some laboratory animals demonstrate carcinogenic activity. Individuals with preexisting diseases of the central nervous system, lungs, liver, or kidneys may have increased susceptibility to the toxicity of excessive exposures. **Ethylene glycol monobutyl ether acetate** Can be absorbed through the skin in harmful amounts. May destroy red blood cells. May cause abnormal kidney function. **Methyl ethyl ketone** High concentrations have caused embryotoxic effects in laboratory animals. Methyl ethyl ketone has been demonstrated to potentiate (i.e., shorten the time of onset) the peripheral neuropathy caused by either n-hexane or methyl n-butyl ketone. MEK by itself has not been demonstrated to cause peripheral neuropathy. Liquid splashes in the eye may result in chemical burns. **Methyl isobutyl ketone** Recurrent overexposure may result in liver and kidney injury. Individuals with preexisting diseases of the central nervous system or lungs may have increased susceptibility to the toxicity of excessive exposures. **Polyester resin B** Contact may cause skin irritation with discomfort or rash. May cause eye irritation with discomfort, tearing, or blurred vision. **Propylene glycol monomethyl ether acetate** May cause moderate eye burning. Recurrent overexposure may result in liver and kidney injury. **Toluene** Recurrent overexposure may result in liver and kidney injury. High airborne levels have produced irregular heart beats in animals and occasional palpitations in humans. Rats exposed to very high airborne levels have exhibited high frequency hearing deficits. The significance of this to man is unknown. **WARNING:** This chemical is known to the State of California to cause birth defects or other reproductive harm. Chromosomal changes in the circulating blood of exposed work been reported. The significance of these reports is unclear of exposure to other substances. Individuals with preexisting diseases of the central nervous system may have increased susceptibility to the toxicity of excessive exposures. **Xylene** Recurrent overexposure may result in liver and kidney injury. Can be absorbed through the skin in harmful amounts. Individuals with pre-existing disease of the central nervous system, kidneys, liver, cardiovascular system, lungs, or bone marrow may have increased susceptibility to the toxicity of excessive exposures. **1,6-hexamethylene diisocyanate** May cause temporary upper respiratory and/or lung irritation with cough, difficult breathing, or shortness of breath. Overexposure may cause asthma-like reactions with shortness of breath, wheezing, cough, which may be permanent; or permanent lung sensitization. This effect may be delayed for several hours after exposure. Prolonged skin contact may cause chemical burns. Liquid splashes in the eye may result in chemical burns. Individuals with preexisting lung disease, asthma or breathing difficulties may have increased susceptibility to the

toxicity of excessive exposures.

Section VI - Reactivity Data

Stability: Stable
Incompatibility (materials to avoid): None reasonably foreseeable.
Hazardous decomposition products: CO, CO_2, smoke.
Hazardous polymerization: Will not occur.

Section VII - Spill or Leak Procedures

Steps to be taken in case material is released or spilled: Do not breathe vapors. Do not get in eyes or on skin. Wear a positive pressure supplied air vapor/particulate respirator NIOSH (TC-19C) eye protection, gloves and protective material. Remove sources of ignition. Absorb with inert material. Ventilate area. Pour liquid decontaminate solution over the spill and allow to sit 10 minutes, minimum. Typical decontamination solutions are:

        20% Surfactant (Tergitol TMN 10)
        80% Water
or   0-10% Ammonia
        2-5% Detergent
        Balance water

Pressure can be generated. <u>Do not</u> seal container. After 48 hours, material may be sealed and disposed of.
Waste disposal method: Do not allow material to contaminate ground water systems. Incinerate absorbed material in accordance with federal, state, and local requirements. Do not incinerate in closed containers.

Section VIII - Special Protection Information

Respiratory: Do not breathe vapors or mists. Wear a positive pressure supplied air respirator NIOSH (TC-19C) while mixing activator with any paint or clear enamel, during application and until all vapors and spray mists are exhausted. Individuals with a history of lung or breathing problems or prior reaction to isocyanate should not use or be exposed to this product. Do not permit anyone without protection in the painting area. Follow the respirator manufacturer's directions for respirator use.
Ventilation: Provide sufficient ventilation in volume and pattern to keep contaminants below applicable exposure limits.
Protective clothing: Neoprene gloves and coveralls are recommended.
Eye protection: Desirable in all industrial situations. Include splash guards or side shields.

Section IX - Special Precautions

Precautions to be taken in handling and storing: Observe label precautions. Keep away from heat, sparks and flame. Close container after each use. Ground containers when pouring. Wash thoroughly after handling and before eating or smoking. Do not store above 120°F.
Other precautions: Do not sand, flame cut, braze or weld dry coating without a NIOSH approved respirator or appropriate ventilation.

Section X - Other Information

Section 313 Supplier Notification: The chemicals listed below with percentages are subject to the reporting requirements of Section 313 of the Emergency Planning and Right-To-Know Act of 1986 and of 40 CFR 372.

PRODUCT CODE   INGREDIENTS (See Section II)

7500S acrylic polymer-C, butyl acetate, ethyl acetate, ethylbenzene (2-6%*), hexyl acetate isomers, methyl isobutyl ketone (3%*), polyester resin, propylene glycol monomethyl ether acetate, toluene (1%*), xylene (18-22%*)
GAL WT: 8.02 WT PCT SOLIDS: 32.94 VOL PCT SOLIDS: 27.13
SOLVENT DENSITY: 7.38 VOC LE: 5.4 VOC AP: 5.4 H: 2 F: 3
R: 0 FLASH PT: BETWEEN 20 - 73 F (CC) OSHA STORAGE: IB

7565S aliphatic polymeric isocyanate-B, ethyl acetate,
GAL WT: 8.38 WT PCT SOLIDS: 47.98 VOL PCT SOLIDS: 41.66
SOLVENT DENSITY: 7.47 VOC LE: 4.4 VOC AP: 4.4 H: 3 F: 3
R: 1 FLASH PT: BETWEEN 20 - 73 F (CC) OSHA STORAGE: IB

FIGURE 1-1 (Continued)

7575S aliphatic polymeric isocyanate-B, butyl acetate, ethyl acetate, ethylbenzene (2-7%*), toluene (4%*), xylene (20-24%*)
GAL WT: 8.26 WT PCT SOLIDS: 48.29 VOL PCT SOLIDS: 41.34
SOLVENT DENSITY: 7.28 VOC LE: 4.3 VOC AP: 4.3 H: 3 F: 3
R: 1 FLASH PT: BETWEEN 20 - 73 F (CC) OSHA STORAGE: IB

7585S aliphatic polymeric isocyanate-B, ethylbenzene (2-6%*), hexyl acetate isomers, propylene glycol monomethyl ether acetate, xylene (17-20%*)
GAL WT: 8.37 WT PCT SOLIDS: 47.72 VOL PCT SOLIDS: 41.38
SOLVENT DENSITY: 7.46 VOC LE: 4.4 VOC AP: 4.4 H: 3 F: 3
R: 1 FLASH PT: BETWEEN 73 - 100 F (CC) OSHA STORAGE: IC

7595S aliphatic polymeric isocyanate-B, ethyl 3-ethoxy propionate, ethylene glycol monobutyl ether acetate (20%*)
GAL WT: 8.59 WT PCT SOLIDS: 46.43 VOL PCT SOLIDS: 41.36
SOLVENT DENSITY: 7.85 VOC LE: 4.6 VOC AP: 4.6 H: 3 F: 2
R: 1 FLASH PT: BETWEEN 140 - 200 F (CC) OSHA STORAGE: IIIA

7600S acrylic polymer-A, butyl acetate, ethylbenzene (2-6%*), methyl ethyl ketone (12%*), methyl isobutyl ketone (7%*), toluene (18%*), xylene (19-23%*)
GAL WT: 7.75 WT PCT SOLIDS: 35.50 VOL PCT SOLIDS: 29.07
SOLVENT DENSITY: 7.05 VOC LE: 5.0 VOC AP: 5.0 H: 2 F: 3
R: 0 FLASH PT: BETWEEN 20 - 73 F (CC) OSHA STORAGE: IB

7601S acrylic polymer, aromatic hydrocarbon, butyl acetate, ethylbenzene (1-4%*), methyl ethyl ketone (28%*), propylene glycol monomethyl ether acetate, toluene (28%*), xylene (12-15%*), 1,2,4-trimethyl benzene (0-2%*)
GAL WT: 7.17 WT PCT SOLIDS: 3.63 VOL PCT SOLIDS: 2.75
SOLVENT DENSITY: 7.11 VOC LE: 6.9 VOC AP: 6.9 H: 2 F: 3
R: 0 FLASH PT: BETWEEN 20 - 73 F (CC) OSHA STORAGE: IB

7655S aliphatic polyisocyanate resin-A, aromatic hydrocarbon, butyl acetate, toluene (27%*),
GAL WT: 8.08 WT PCT SOLIDS: 39.00 VOL PCT SOLIDS: 32.41
SOLVENT DENSITY: 7.29 VOC LE: 4.9 VOC AP: 4.9 H: 3 F: 3
R: 1 FLASH PT: BETWEEN 20 - 73 F (CC) OSHA STORAGE: IB

7675S aliphatic polyisocyanate resin-A, aromatic hydrocarbon, butyl acetate, propylene glycol monomethyl ether acetate
GAL WT: 8.32 WT PCT SOLIDS: 37.50 VOL PCT SOLIDS: 32.06
SOLVENT DENSITY: 7.65 VOC LE: 5.2 VOC AP: 5.2 H: 3 F: 3
R: 1 FLASH PT: BETWEEN 73 - 100 F (CC) OSHA STORAGE: IC

7695S aliphatic polyisocyanate resin-A, aromatic hydrocarbon, butyl acetate, cumene (0-1%*), ethyl 3-ethoxy propionate, 1,2,4-trimethyl benzene (1-7%*)
GAL WT: 8.33 WT PCT SOLIDS: 37.48 VOL PCT SOLIDS: 32.11
SOLVENT DENSITY: 7.67 VOC LE: 5.2 VOC AP: 5.2 H: 3 F: 2
R: 1 FLASH PT: BETWEEN 100 - 140 F (CC) OSHA STORAGE: II

7800S acrylic polymer-B, aromatic hydrocarbon, butyl acetate, cumene (0-1%*), ethyl acetate, methyl ethyl ketone (12%*), polyester resin, propylene glycol monomethyl ether acetate, toluene (2%*), xylene (0-1%*), 1,2,4-trimethyl benzene (1-5%*)
GAL WT: 8.06 WT PCT SOLIDS: 40.52 VOL PCT SOLIDS: 33.85
SOLVENT DENSITY: 7.25 VOC LE: 4.8 VOC AP: 4.8 H: 2 F: 3
R: 0 FLASH PT: BETWEEN 20 - 73 F (CC) OSHA STORAGE: IB

7875S aliphatic polymeric isocyanate-B, ethyl acetate, toluene (8%*), 1,6-hexamethylene diisocyanate (<0.5%*),
GAL WT: 8.89 WT PCT SOLIDS: 71.33 VOL PCT SOLIDS: 65.69
SOLVENT DENSITY: 7.43 VOC LE: 2.5 VOC AP: 2.5 H: 3 F: 3
R: 1 FLASH PT: BETWEEN 20 - 73 F (CC) OSHA STORAGE: IB

7895S aliphatic polymeric isocyanate-B, ethyl 3-ethoxy propionate, ethylene glycol monobutyl ether acetate (5%*), 1,6-hexamethylene diisocyanate (<0.5%*)
GAL WT: 9.02 WT PCT SOLIDS: 69.17 VOL PCT SOLIDS: 64.66
SOLVENT DENSITY: 7.87 VOC LE: 2.8 VOC AP: 2.8 H: 3 F: 2
R: 1 FLASH PT: BETWEEN 100 - 140 F (CC) OSHA STORAGE: II

7899S acetic acid ester of c9-11 oxo-alcohol, aliphatic polymeric isocyanate-B, hexyl acetate isomers, 1,6-hexamethylene diisocyanate (<0.5%*)
GAL WT: 8.81 WT PCT SOLIDS: 70.92 VOL PCT SOLIDS: 64.74
SOLVENT DENSITY: 7.27 VOC LE: 2.6 VOC AP: 2.6 H: 3 F: 1
R: 1 FLASH PT: BETWEEN 100 - 140 F (CC) OSHA STORAGE: II

Notice: The data in this material safety data sheet relate only to the specific material designated herein and do not relate to use in combination with any other material or in any process.

Product Manager - Refinish Sales

Prepared by D. G. Detweiler

19-11

FIGURE 1–1 (Continued)

FIGURE 1–2 *This is a supplied-air respirator hood. Fresh, uncontaminated air is pumped into the hood.*

inhaling **isocyanate** paint vapors and mists as well as from hazardous solvent vapors.

One type of supplied air respirator system consists of a hood (Figure 1–2) or a mask, an air hose and a separate oil-less air pump (Figure 1–3). The pump is located in an area where it can obtain uncontaminated air (always outside of the spray booth). The clean air is pumped into the hood or mask. This way the painter does not have to breathe contaminated air from the spray booth. The hood has an advantage over the mask in that the hood also protects the head (especially the eyes) from contact with isocyanates. The hood lens can be protected from overspray contamination with a clear, disposable shield. The hood or mask should be cleaned with soap and water after each use.

Another type of supplied air respirator system uses air from the shop's air compressor. In this system the shop air compressor must be located where it can obtain fresh air. An inline carbon monoxide monitor and alarm is installed on the air line. If the air compressor were to overheat, the heat may convert to

FIGURE 1–3   *This is a supplied-air respirator pump. It sits outside of the spray booth and provides air to the supplied-air respirator.*

FIGURE 1–4   *These are replaceable organic vapor cartridges. The one on the right is exposed to air and will have a finite life. The one on the left is in an airtight bag.*

hydrocarbons in the compressor lubricating oil to carbon monoxide—a deadly odorless gas. The monitor detects the presence of carbon monoxide, and the alarm alerts the painter to danger. The compressed air is filtered to remove dust, oil, and water. Usually these filters are mounted on the spray booth wall. An activated carbon filter mounted on the wall or on the painter's belt removes organic odors and oil vapor. This system can use either a hood or a mask.

Cartridge Filter Respirators

If the refinishing system that is sprayed contains no isocyanates, an air-purifying, cartridge respirator with organic vapor cartridges (Figure 1–4) and prefilters can be used. These respirators protect against vapors and spray mists of nonactivated enamels, and other nonisocyanate materials.

This type of respirator consists of a rubber face piece designed to conform to the face and form an airtight seal. It includes replaceable prefilters and cartridges that remove solvent and other vapors from the air. The paint respirator also has intake and exhaust valves, which ensure that all incoming air flows through the filters.

It is very important with air-purifying cartridge respirators that they fit securely around the edge of the face to prevent contaminated air from leaking into the breathing area. To check this, a qualitative fit test should be done prior to using the respirator, performing both negative and positive pressure checks. To check for negative pressure, the wearers should place

FIGURE 1–5   *This painter wears a disposable cartridge respirator. When the organic vapor cartridge is used up, the entire mask is discarded.*

the palms of their hands over the cartridges and inhale. A good fit will be evident if the face piece collapses onto the wearer's face. To perform a positive pressure check, the wearer covers up the exhalation valve and exhales. A proper fit is evident if the face piece billows out without air escaping the mask. Another form of quantitative fit testing consists of exposing amyl acetate (banana oil) near the seal around the face. If no odor is detected, a proper fit is evident.

Cartridge respirators are available in several sizes and might or might not contain a face mask (Figure 1-5). The most common size will provide the best protection. But, wearers of this type of respirator should be aware that facial hair might prevent an airtight seal, presenting a hazard to the wearer's health. Therefore, refinishers with facial hair should use a positive pressure-supplied air respirator system, because hair will prevent a seal of mask to face, eliminating the respirator's effectiveness. Remember that cartridge respirators should be used only in well-ventilated areas. They must not be used in environments containing less than 19.5 percent oxygen.

To maintain the cartridge filter respirator, keep it clean and change the prefilters and cartridges as often as directed by the manufacturer. Here are a few other maintenance tips:

- Replace the prefilters when it becomes difficult to breathe through the respirator.
- Replace the cartridge after 8 to 10 hours of use.
- Regularly check the mask to make sure it does not have any cracks or dents.
- Store the respirator in an airtight container. Exposure to air will shorten the useful life of the cartridge.
- Follow the manufacturer's instructions provided to ensure proper maintenance and fit.

WARNING: The precautionary measures mentioned here should prevent overexposure by inhalation, but if any symptoms of overexposure develop (such as breathing difficulty, tightness of chest, severe coughing, irritation of the nose or throat, or nausea), leave the area quickly and get fresh air. If breathing continues to be labored, see a physician immediately.

It is important to note that symptoms of inhalation overexposure might not appear until 4 to 8 hours after the exposure. Depending upon the severity of overexposure, symptoms might persist for 3 to 7 days.

Dust or Particle Respirators

To protect against dust from sanding, use a dust respirator (Figure 1–6). Sanding operations in the collision repair facility create dust that can cause bronchial irritation and possible long-term lung damage if inhaled. Protection from this health hazard is necessary; a NIOSH-approved dust respirator should be worn whenever a refinisher or someone working close to him or her is involved in a sanding operation. Follow the instructions provided with the dust respirator to ensure proper maintenance fit. Remember that dust masks do not protect against vapors and spray mists.

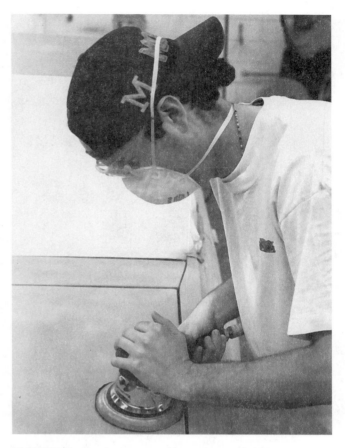

FIGURE 1–6 A dust respirator (two straps) should be used when dry sanding.

HEAD PROTECTION

Be sure to tie long hair securely behind the head before beginning to work on a vehicle. The hair also must be protected against dust and sprays. To keep hair clean (and healthy) wear a cap at all times in the work area and a protective painter's stretch hood in the spray booth.

Eye Protection

The eyes are sensitive to dust and flying particles from grinding or hammering to vapors from spraying, and to liquid solvents when mixing paint. Such exposure could cause severe pain and possibly the loss of sight. Whenever such risks exist, a full-face shield, a good pair of safety glasses (Figure 1–7) or a supplied air respirator hood should be worn. Remember that eyes are irreplaceable. If solvent splashes into someone's eye, first check the eye for a contact lens. Remove the contact lens if present. Flush the eye immediately with water for at least 15 minutes. See a

FIGURE 1–7 *This technician wears safety glasses. The minimum requirement should be worn at all times in the shop.*

FIGURE 1–8 *This painter is wearing the following safety apparel: disposable jumpsuit, gloves, disposable respirator, spray sock, and safety glasses.*

physician immediately after flushing, even if no irritation or redness is present.

Ear Protection

Repeated exposure to loud noises—air chiseling, hammering, or grinding—may result in gradual hearing loss to unprotected ears. Ear protection in the form of ear plugs or muffs will minimize the chance of hearing loss.

BODY PROTECTION

Loose clothing, unbuttoned shirtsleeves, dangling ties, loose jewelry, and shirts hanging out are *very* dangerous in a collision repair facility. Instead, wear approved shop coveralls or jumpsuits (Figure 1–8). Pants should be long enough to cover the tops of the shoes. Keep them mended and tear-free.

A clean jumpsuit or lint-free coveralls should be worn when in the spray area. Dirty, solvent-soaked clothing will hold these chemicals against the skin, causing irritation or a rash. Make sure they are long-sleeved for complete protection.

SKIN PROTECTION

Paint vapor, liquid solvents, and the hazardous chemicals they contain can penetrate the skin. Although skin penetration of vapor and solvent is not as fast as breathing solvent vapor, hazardous chemicals can reach the bloodstream.

The harmful effects of vapors and solvents on the skin can be prevented very effectively by wearing proper gloves (Figure 1–8). Impervious gloves, such as the nitrile latex type, should be used when working with vapors and solvents. These gloves offer special protection from the materials found in paint products.

TABLE 1–1: GLOVES

Features	Advantages	Benefits
Synthetic latex material	1. Hazard resistant 2. Flexible	1. Helps resist abrasion, cuts, snags, and punctures. 2. Excellent dexterity to perform even delicate tasks
17 mils thick, 11 inches long, case hardened	Greater wear resistance and tensile strength	Longer lasting
Flock lined	Soft fit	Wearer comfort and absorbs perspiration

See the MSDS for glove recommendations. Thick, strong work gloves should be worn in the prep area to avoid cuts or abrasions. (Table 1-1 gives the advantages and benefits of various glove types.) Always remember to wash hands thoroughly when leaving the shop area. This provides protection from ingesting any harmful elements that may have been touched. Cover as much skin as possible when spraying paint.

Hands should be cleaned with a proper hand cleaner. At the end of a day's work, it is wise to oil the skin a little by applying a good silicone-free skin cream. Do not use thinner as a hand cleaner.

WARNING: Do not use solvent as a hand cleaner. Lacquer thinner used to clean hands can enter the bloodstream as well as cause a rash or dermatitis.

FOOT PROTECTION

Wear safety work shoes that have metal toe inserts and nonslip soles. The inserts protect the toes from falling objects; the soles help to prevent falls. In addition, good work shoes provide support and comfort for someone who is standing for a long time. *Never* wear sneakers, gym shoes, or dress shoes; none of these shoes provide adequate protection in a refinishing shop.

When spraying, many painters wear disposable shoe covers. In fact, disposable garments and hoods are becoming more commonly used by sprayers.

DAY-BY-DAY PERSONAL SAFETY GUIDE

The following guidelines are designed to protect the painter while on the job, from the moment he or she determines what steps need to be taken and the products to be used until the time comes to put away the equipment and get ready to head home.

- **Be informed.** Read the warnings on the product labels and in manufacturers' literature. If more information is desired, get copies of the Material Safety Data Sheets for specific products from the shop's office or from the paint suppliers. As previously mentioned, these contain information on hazardous ingredients and protective measures that the painter should use.
- **During power sanding.** When one is power sanding, dust and dirt fly into the air. These can get into eyes, lungs, and scalp without proper protection. Safety glasses or goggles will protect the eyes. Do not wear contact lenses when grinding, sanding, or handling **solvents.** Head covers provide scalp and hair protection. A NIOSH-approved dust particle mask should also be worn to avoid inhaling dust and particles. All masks should fit tightly to your skin.
- **During cleaning with compressed air.** When using a dust gun to clean doorjambs and other hard-to-reach places, eye protection, ear protection and particle masks should always be worn.
- **During metal conditioning.** Metal conditioners contain phosphoric acid. Breathing these chemicals or allowing them to come in direct contact with the skin, eyes, or clothing may cause irritation. The use of safety glasses (to prevent splashes into the eyes), coveralls, rubber gloves, and a NIOSH-approved organic vapor respirator is recommended when using these products. If the coveralls become soaked for any reason, make sure they are changed to clean ones to avoid irritation to the skin; or soak with water to dilute the chemicals.
- **During mixing and handling.** Mixing and pouring of refinish materials should be done in a well-ventilated location away from areas used to store or apply the products. When opening cans or mixing, materials might splash. To avoid splashes to the eyes, goggles or other protection should be worn.
- **During spraying of undercoats or topcoats.** The application of undercoats and topcoats requires the use of spraying equipment, which

FIGURE 1–9   *Paint cabinet.*

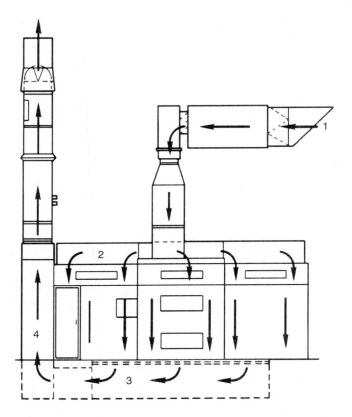

FIGURE 1–10   *Air makeup unit brings in outside air, heats and filters it, and then pushes the air through the roof of the downdraft booth. The exhaust fan pulls the contaminated spray booth air through filters and then outside.*

can be hazardous if not used properly. Painters should be fully protected when applying undercoat or topcoat products. Coveralls, rubber gloves, safety shoes, eye protection, and an appropriate respirator and/or protective hood for the product being applied should be worn throughout the process.

- **Storing paint materials.** All refinish products should be stored away from the actual work area. Paint kept in the work area should be limited to a one-day supply. Empty containers should be disposed of daily. All partially used containers should be kept securely closed and should be placed in proper metal (fire-resistant) storage cabinets (Figure 1–9) at the end of the day.
- **Before leaving the shop.** Solvents, chemicals, and other materials can contaminate clothing and wind up on the hands when removing personal protective equipment or putting away the refinishing tools. They can still enter one's system through the body's digestive tract if the hands are not washed before eating, drinking, smoking, or using toilet facilities.

 **GENERAL SHOP PROCEDURES**

In addition to personal safety, the refinishing technician must be aware of general safety procedures. The following are some of the rules and precautions that should be observed to insure a safe and healthful work environment in the paint shop.

ENVIRONMENTAL CONTROLS

Persons working in body/paint shop facilities are often exposed to dangerous amounts of various gases, dusts, and vapors. Because of this exposure, control measures should be established and practiced for the following frequently observed air contaminants and other hazardous substances:

- **Ventilation.** Proper ventilation is very important in areas where caustics, degreasers, undercoats, and finishes are used. In the paint shop and in the area where vehicles are prepared, ventilation can be by means of an air-exchange system, extraction floors, or central dust extraction, combined with good extraction power for spraying area walls. For the spray booth (Figure 1–10), adequate air replacement is necessary not only to promote evaporation and drying of the

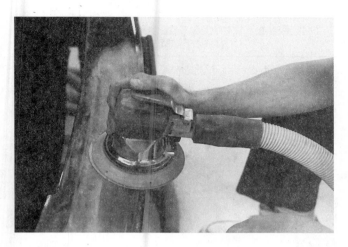

FIGURE 1–11 _Central vacuum system in use. This dual-action sander has holes in the sandpaper and pad._

areas sprayed, but also to remove harmful mist and vapors.

- **Carbon monoxide.** Operate the engine only in a well-ventilated area to avoid the danger of carbon monoxide (**CO**). If the shop is equipped with a tailpipe exhaust system to remove CO from the garage, use it. If not, use direct piping to the outside or a mechanical ventilation system.
- **Paints, body fillers, and thinners.** Thinners used in most paints have a narcotic effect, and long-term exposure can eventually cause irreparable damage. In addition to the ventilation in the spray area or paint booth, respirators should be worn. As previously mentioned, rubber or safety gloves should be worn while handling paints and thinners. If any of these materials get on the skin, promptly wash the affected area with soap and water.
- **Dust.** Dust is a problem in paint shops. It is produced during operations such as sanding paints, primers, and body fillers. When doing this type of work, wear a dust particle respirator or mask.

Many shops are installing so-called dustless sanding machine systems. Depending on the system, vacuum pumps, vacuum pullers, brush motors, or turbine motors can be used, all in the quest for sufficient air volume and/or velocity to pull airborne sanding dust through either holes in a special sanding pad or a shroud that entirely surrounds the sanding pad. Some systems run constantly; others are started on-demand by plugging a vacuum hose into the vacuum outlet or by pushing a button at the sanding tool end of the hose. Figure 1–11 shows a vacuum system.

By using a dustless system, the paint shop can comply with OSHA's airborne dust standards and eliminate costly, nonproductive cleanups. Dustless systems can be quite efficient at removing the toxic dust created by sanding lead- or chrome-based automotive paint and primer, which can contaminate the work area.

1.4 — ENVIRONMENTAL PROTECTION

HAZARDOUS WASTE

Hazardous wastes are generated by all collision repair facilities. Substances identified as hazardous waste are classified into one or more of the following categories:

1. **Ignitable.** Produces a vapor that if ignited will cause a fire
2. **Corrosive.** Dangerous _pH_ level, below 2 or above 12.5, can cause acid or alkali burns to the skin
3. **Reactive.** Can explode or generate toxic fumes when mixed
4. **Toxic.** Poisonous materials

Common collision repair facility hazardous wastes include:

- **Dirty cleaning solvent.** Thinner used to clean paint equipment
- **Paint aqueous waste.** Wash water from sanding and water-based paint waste
- **Paint organic waste.** Solvent-based paint mixed but not sprayed on the vehicle
- **Hazardous trash.** Material from cleanup of spilled paint
- **Spent filter media.** Used spray booth filters
- **Solid paint waste.** Dried, unused paint

Due to the amount of hazardous wastes produced, most collision repair facilities are identified as conditionally exempt small quantity generators by the Environmental Protection Agency (**EPA**). Records must be kept on how the waste is stored. The generator— the shop—is responsible for proper waste disposal, even after the waste is removed from the shop by an authorized hazardous waste hauler. This is known as cradle-to-grave responsibility. Each shop is required to properly dispose of all hazardous waste. Paint organic waste is kept in a labeled, dated container (Figure 1–12) until removal by an authorized hazardous waste hauler (Figure 1–13). The hauler takes the waste for disposal, for incineration, or recycling.

FIGURE 1–12 Waste thinner barrel in a collision repair facility.

FIGURE 1–13 Waste thinner removed by an authorized hazardous waste hauler.

TABLE 1–2:	NATIONAL REGULATIONS FOR VOC CONTENT LIMITS "AS APPLIED"
Pretreat./wash primer	6.5 VOC/lbs/gal.
Surfacer	4.8 VOC/lbs/gal.
Sealer	4.6 VOC/lbs/gal.
S.S. & B/C topcoat	5.0 VOC/lbs/gal.
Tri & quad coats	5.2 VOC/lbs/gal.
Specialty coating	7.0 VOC/lbs/gal.

FIGURE 1–14 VOC label on a Du Pont primer-sealer.

AIR POLLUTION

For paint to be sprayed, it must be mixed with a solvent. The solvent dilutes the paint to a sprayable concentration. This solvent evaporates into the air during spraying and as the paint dries. It is not trapped by the spray booth filters. The evaporating solvent is called volatile organic compound or **VOC.** VOCs released into the atmosphere cause photochemical smog. The EPA has set up regulations for the allowable amount of VOCs in paint (Table 1-2). The VOC amounts are measured in pounds per gallon as applied. All paints sold must be within the established limits (Figure 1–14). To meet some of these requirements, paint manufacturers have developed many high solids paint products. When compared to previous paint products, these new products have more solids and less solvent. Less solvent causes less VOCs to be released when the new products are sprayed. Some areas of the United States may have stricter regulations than the national rule. These areas may require:

- Special spray guns
- Record keeping
- Spray booth regulations
- Inspections

VEHICLE HANDLING IN THE SHOP

When handling a vehicle in the shop, keep the following safety precautions in mind:

- Set the parking brake when working on the vehicle. If the car has an automatic transmission, set it in PARK unless instructed otherwise for a specific service operation. If the vehicle has a manual transmission, it should be in REVERSE (engine off) or NEUTRAL (engine on).
- If for some reason a work procedure requires working under a vehicle, use safety stands.
- To prevent serious burns, avoid contact with hot metal parts such as the radiator, exhaust manifold, tailpipe, catalytic converter, and muffler.
- Keep clothing and oneself clear from moving parts when the engine is running, especially the radiator fan blades and belts.
- Be sure that the ignition switch is always in the OFF position, unless otherwise required by the procedure.
- When moving a vehicle around the shop, be sure to look in all directions and make certain that nothing is in the way.

HORSEPLAY

No horseplay, running, or fooling around should be attempted in any shop. One thing can lead to another and eventually cause an injury. Horseplay is distracting and wastes time.

HANDLING OF SOLVENT AND OTHER FLAMMABLE LIQUIDS

The refinisher will be working with various solvents to clean surfaces and equipment and to thin finishes. These solvents are extremely flammable. Fumes in particular can ignite explosively.

The following safety practices will help avoid fire and explosion:

- Do not light matches or smoke in the spraying and paint area, and make sure that the hands and clothing are free from solvent when lighting

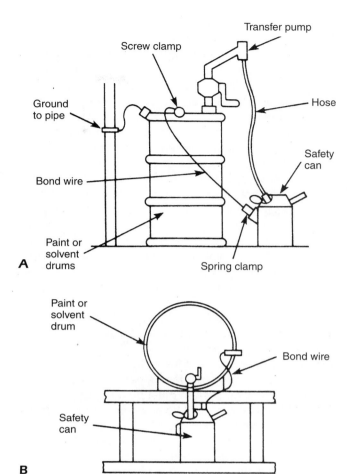

FIGURE 1–15 _Two safe methods of moving flammable liquids from a drum to a portable safety can._

matches or smoking in other areas of the shop where smoking or an open flame is permitted.
- All ignition sources should be carefully controlled and monitored to avoid any possible fire hazard where a high concentration of vapor from highly flammable liquids might at times be present.
- A UL (Underwriters Laboratories) approved drum transfer pump/drum pump along with a drum vent should be used when working with drums to transfer chemicals.
- Keep all solvent containers closed, except when pouring.
- Handle all solvents (or any liquids) with care to avoid spillage. Extra caution should also be used when transferring flammable materials from bulk storage. The most important thing to remember is to make sure the drum is grounded and that a bond wire connects the drum to a safety can (Figure 1–15). Otherwise, static electricity can

build up enough to create a spark that could cause an explosion.

- Discard or clean all empty solvent containers as prescribed by local regulations. Solvent fumes in the bottom of these containers are prime ignition sources. Remember: Never use gasoline as a cleaning solvent.
- Paints, thinners, solvents, and other combustible materials used in the body and paint shop must be stored in approved and designated metal (never wood) storage cabinets or rooms. Storage rooms should have adequate ventilation, which takes harmful fumes and pollutants away from the actual working area. The storage room may require a specific air exchange rate; check local code. Many paint shops use a separate facility for the bulk storage of flammable material. Never have more than one day's supply of paint outside of approved storage areas.
- The connectors on all drums and pipes of flammable and combustible liquids must be vapor- and liquid-tight.
- When spraying paint follow these procedures:
  — Remove portable lamps before spraying.
  — Ventilation system must be turned on.
  — Spray areas must be free from hot surfaces such as heat lamps.
  — The spray area must be kept clean of combustible residue.
  — Ventilation system must be left on while the paint is drying.

FIRE PROTECTION

Every paint shop requires fire extinguishers (Figure 1–16). Since fires are classified as Class A, B, C, and D type, there are different types of extinguishers specially designed for a particular class of fire. Table 1-3 gives the common classes of fire that are found in paint shops and methods of containing them. Some extinguishers are capable of fighting more than one type of fire.

However, the mere provision of a fire extinguisher is useless unless those who might come in contact with it know how to use it properly. If a fire breaks out, there is no time to lose figuring out how to use the fire extinguisher effectively. Operating instructions are imprinted on each listed or approved extinguisher. The approval agencies require information on the front of extinguishers indicating their classification, the relative extinguishing effectiveness (the numeral preceding the classification letter), and the methods of use. However, during an emergency there might be no time to read the label. The basic information should be known ahead of time by anyone who might come in contact with and need to use the fire extinguisher.

FIGURE 1–16 *All collision repair and paint areas should have fire extinguishers capable of fighting class A, B, and C fires.*

A fire can be extinguished by depriving it of its essential ingredients, which are heat, fuel, and oxygen. Most extinguishers work by cooling the fire and removing the oxygen. If the fire extinguisher is going to be used effectively, it must be aimed at the base of the flame where the fuel is located. Fire extinguishers should be checked regularly and be placed at strategic shop locations.

GOOD HOUSEKEEPING

Here are some simple good housekeeping precautions that often go unattended:

- Keep aisles and walkways free of tools, creepers, and any material that might cause a person or fellow worker to trip or stumble.
- Keep floors clean. Oil, paint, or other materials that are spilled should be cleaned up immediately. Sand, earth, or absorbents can be used to absorb the liquids. But once they have done their job, they should also be cleaned up. Make sure that these liquids are not discharged through the floor drains or other outlets leading to public waterways.
- Any dirty rags or other combustible material must be deposited in a metal container with a suitable metal cover and should be removed to a safe place outside the building. Keep used paper towels and other paper products in a separate, covered container, which should be emptied every day.
- Customers and all nonemployees should never be allowed in any of the paint refinishing shop's work areas.

Personal protective equipment, a properly maintained paint shop environment, and attention to good

TABLE 1–3:   GUIDE TO EXTINGUISHER SELECTION

	Class of Fire	Typical Fuel Involved	Type of Extinguisher
Class A Fires (green)	**For Ordinary Combustibles** Put out a Class A fire by lowering its temperature or by coating the burning combustibles.	Wood Paper Cloth Rubber Plastics Rubbish Upholstery	Water*[1] Foam* Multipurpose dry chemical[4]
Class B Fires (red)	**For Flammable Liquids** Put out a Class B fire by smothering it. Use an extinguisher that gives a blanketing, flame-interrupting effect; cover whole flaming liquid surface.	Gasoline Oil Grease Paint Lighter fluid	Foam* Carbon dioxide[5] Halogenated agent[6] Standard dry chemical[2] Purple K dry chemical[3] Multipurpose dry chemical[4]
Class C Fires (blue)	**For Electrical Equipment** Put out a Class C fire by shutting off power as quickly as possible and by always using a nonconducting extinguishing agent to prevent electric shock.	Motors Appliances Wiring Fuse boxes Switchboards	Carbon dioxide[5] Halogenated agent[6] Standard dry chemical[2] Purple K dry chemical[3] Multipurpose dry chemical[4]
Class D Fires (yellow)	**For Combustible Metals** Put out a Class D fire of metal chips, turnings, or shavings by smothering or coating with a specially designed extinguishing agent.	Aluminum Magnesium Potassium Sodium Titanium Zirconium	Dry powder extinguishers and agents only

*Cartridge-operated water, foam, and soda-acid types of extinguishers are no longer manufactured. These extinguishers should be removed from service when they come due for their next hydrostatic pressure test.

Notes:

(1) Freezes in low temperatures unless treated with antifreeze solution, usually weighs over 20 pounds, and is heavier than any other extinguisher mentioned.

(2) Also called ordinary or regular dry chemical. (sodium bicarbonate)

(3) Has the greatest initial fire-stopping power of the extinguishers mentioned for Class B fires. Be sure to clean residue immediately after using the extinguisher so sprayed surfaces will not be damaged. (potassium bicarbonate)

(4) The only extinguishers that fight A, B, and C classes of fires. However, they should not be used on fires in liquefied fat or oil of appreciable depth. Be sure to clean residue immediately after using the extinguisher so sprayed surfaces will not be damaged. (ammonium phosphates)

(5) Use with caution in unventilated, confined spaces.

(6) May cause injury to the operator if the extinguishing agent (a gas) or the gases produced when the agent is applied to a fire are inhaled.

safety and health practices all play important roles in the refinisher's good health. Taking the time to properly prepare *before* working on a vehicle can prevent many accidents or potentially dangerous chemical exposures. Repetition of careful safety procedures will turn them into habits—good habits that will contribute to a long, healthy life.

1.5 — ASE CERTIFICATION

Just as doctors, nurses, accountants, electricians, and other professionals are licensed or certified to practice their professions, a painter can also be

---

TABLE 1–4:   PAINTING AND REFINISHING

Content Area	Number of Questions in Test
A. Surface preparation	12
B. Spray gun operation	7
C. Paint mixing, matching, and applying	13
D. Solving paint application problems	8
E. Finish defects, causes, and cures	6
F. Safety precautions and miscellaneous	4
Total	50

---

certified. Certification protects the general public and the practitioner or professional. It assures the general public and the prospective employer that certain minimum standards of performance have been met. Many employers now expect their refinishing technicians to be certified. The certified technician is recognized as a professional by employers, peers, and the public. For this reason, the certified refinisher usually receives higher pay than the noncertified refinisher.

Refinishers can become certified in one or more technical areas by taking and passing a refinisher certification test. The National Institute for Automotive Service Excellence (ASE) offers a voluntary certification program that is recommended by the major vehicle manufacturers in the United States. The Painting and Refinishing Test contains fifty questions in the areas noted in Table 1–4.

Craftsmen passing the written tests are awarded a certificate, a shoulder emblem, and other credentials attesting to their know-how. ASE certification is a badge of *proven* professionalism.

To help you prepare for the Painting and Refinishing Performance programs, some test questions at the end of each chapter are similar in design and content to those used by the ASE. For further information on the ASE certification program, write: National Institute for Automotive Service Excellence, 1920 Association Drive, Reston, Virginia 22091.

REVIEW QUESTIONS

1. Technician A says that cartridge filter respirators protect the user from isocyanates. Technician B says that when isocyanates are sprayed, a supplied air respirator must be worn. Who is right?
   a. technician A
   b. technician B
   c. both technician A and technician B
   d. neither technician A nor technician B

2. Eye protection should be worn when using
   a. grinders.
   b. disk sanders.
   c. air chisels.
   d. all of the above.

3. Technician A says that one day's supply of paint may be left outside of approved storage. Technician B says that the amount of paint that may be left out of approved storage is 5 quarts maximum. Who is right?
   a. technician A
   b. technician B
   c. both technician A and technician B
   d. neither technician A nor technician B

4. Class C fires are from combustible metals.
   a. True
   b. False

5. Technician A wears gloves when mixing paint. Technician B wears gloves when spraying paint. Who is right?
   a. technician A
   b. technician B
   c. both technician A and technician B
   d. neither technician A nor technician B

6. Information on paint hazards can be found on _____ sheets.

7. Technician A says that the pH scale measures acid or alkaline. Technician B says that alkali can cause burns. Who is right?
   a. technician A
   b. technician B
   c. both technician A and technician B
   d. neither technician A nor technician B

8. Which of the following are hazardous wastes?
   a. gun cleaning solvent
   b. dried paint
   c. spray booth exhaust filters
   d. all of the above

9. Technician A says that VOCs are released when solvent-based paint is sprayed. Technician B says that VOCs cause smog. Who is right?
   a. technician A
   b. technician B
   c. both technician A and technician B
   d. neither technician A nor technician B

10. Which type of extinguisher can be used on all classes of fires?
    a. water
    b. foam
    c. multi-purpose dry chemical
    d. none of the above

Chapter

2

Refinishing Equipment and Its Use

Objectives

After studying this chapter, you will be able to:
- Explain the operation of spray booths, drying rooms, and prep stations.
- Identify air-powered tools in the paint shop.
- Name the electric power tools most commonly used in paint shops.
- Maintain shop power equipment tools.
- Name various other tools and materials used in the refinishing shop.

Key Term List

spray booth
crossdraft
semi-down draft
downdraft
overspray
manometer
cure
spray gun viscosity
metallics
clear coat

To do a good refinishing job, the proper equipment must be used correctly. But to achieve high-quality finishes on automobiles, good materials, proper equipment, and correct techniques are necessities.

The **spray booth,** various air- or electric-powered tools, and other paint shop equipment and materials are presented in this chapter. Chapter 3 presents the spray gun and its use. The selection and use of air compressors, air control equipment, and air hose connectors are thoroughly described in Chapter 4.

2.1 — SPRAY BOOTHS

The collision repair facility, by necessity, is continually generating dust and dirt from the grinding of welds, sanding of filler, and similar dirt-creating operations—the very worst kind of environment in which to paint cars. Much of this dust is so fine it can scarcely be controlled.

Providing a clean, safe, well-illuminated, healthful enclosure for painting is the primary purpose of a spray booth (Figure 2–1). It isolates the painting operation from the dirt- and dust-producing activities and confines and exhausts the volatile fumes created by spraying automotive finishes. Modern spray booths are scientifically designed to create the proper air movement, provide necessary lighting, and enclose the painting operation safely. In addition, their construction and performance must conform to federal, state, and even local safety codes, not to mention

Figure 2–1 *Painting in a spray booth*

those of insurance underwriters. In some areas, automatically operated fire extinguishers are required because of the highly explosive nature of the refinishing materials.

The spray booth should be located as far as possible from the area where dust and dirt are prevalent. Therefore, it should be isolated from the mechanical and metalworking portions of the shop wherever possible. This can be accomplished with partitions, walls, or a separate building arrangement. Paint storage should be outside the booth, but nearby.

If the volume of paint work is sufficient, a straight-line work flow is recommended (Figure 2–2). Utilizing a drive-through–type spray booth, the layout is designed for maximum efficiency of labor and equipment. Jobs are started in the metalworking stalls in the normal manner. From this point the work flows in a production line manner through each of the various stages all the way to final cleanup. The drive-through principle can be used in a one- or two-booth arrangement.

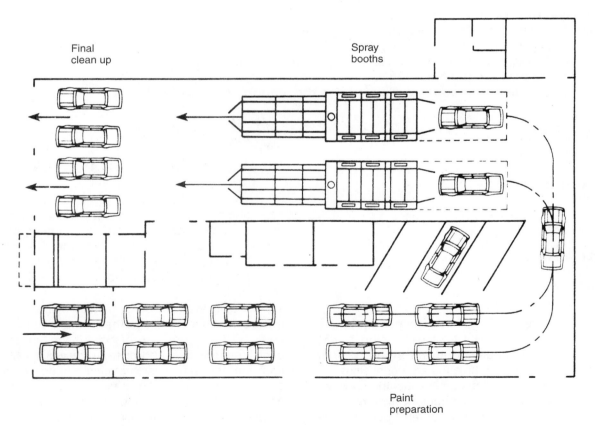

Final clean up

Spray booths

Paint preparation

FIGURE 2–2 *Typical body/paint shop layout showing straight-line work flow finishing operation. The important stops in such an arrangement are: paint preparation, spray booth, and final cleanup. (Courtesy of Du Pont Co.)*

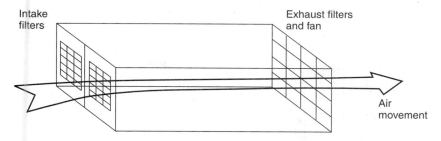

FIGURE 2–3 A negative pressure, exhaust-fan-only booth.

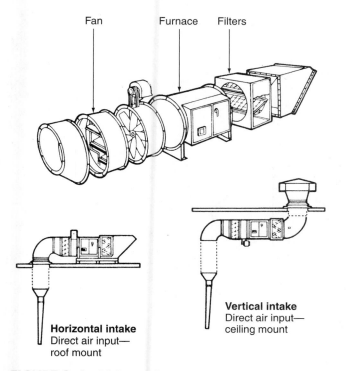

FIGURE 2–4 Makeup air systems.

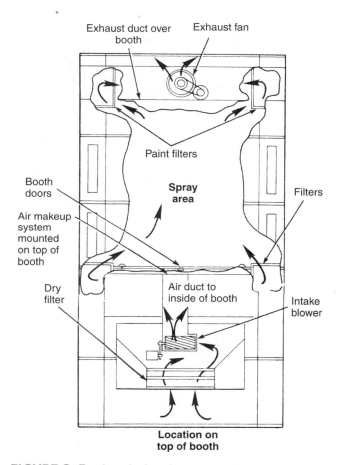

FIGURE 2–5 A typical makeup system.

Spray booths function by moving air. All spray booths have some type of air intake system and air exhaust system. In the simplest system there is an exhaust fan only. In this operation, the intake air from the door or roof is filtered as it is pulled toward the exhaust fan. The air current picks up the paint **overspray;** then it is filtered before it exits the booth (Figure 2–3). Usually the contaminated air is discharged outside the building. This system has negative pressure. An advantage of this system is low cost. But there are serious drawbacks. The intake air will be pulled from the shop and may have dust particles in it, if sanding operations are nearby. Because the system is under negative pressure, if the booth is not airtight, air and dirt will be pulled in from any opening, not just the intake filters. Any door gaps or panel gaps will allow dirty air in. Lastly, in cold climates, as the

shop air is pulled through the booth, the shop heater may not be able to keep the shop warm. Not only will the shop be cold, but the air entering the booth will be cold. This situation can lead to numerous paint problems.

Air makeup systems are used to prevent most of these problems (Figure 2–4 and Figure 2–5). An air makeup system consists of an intake fan, filters and ductwork to bring in air from outside the shop. Some systems have a furnace to provide heated intake air. This system has positive pressure. There is slightly more air pushed into the booth than is exhausted.

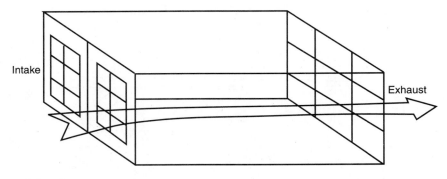

FIGURE 2-6 *Crossdraft booth.*

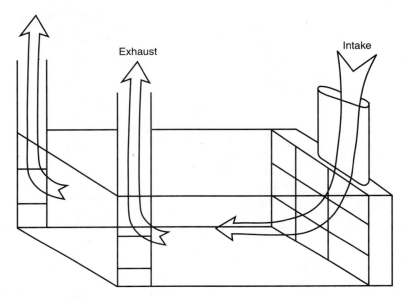

FIGURE 2-7 *Reverse-flow crossdraft booth.*

With this positive pressure, if there are any leaks in the walls or doors, the air will be forced out rather than be pulled in. Heated intake air maintains proper spray booth temperature in cold weather.

The air flows through the booth in one of the following ways:

- Crossdraft
- Semi-downdraft
- Downdraft

A **crossdraft** booth moves the air in from one end of the booth and out through the other end. In a regular flow crossdraft booth, the air enters through filters located in the drive-through doors. The air travels over and around the vehicle and exits through the opposite end (Figure 2–6). In a reverse flow crossdraft booth, the air enters at the solid end of the booth and exits at filters located near the drive-through doors (Figure 2–7). A reverse flow booth can have an air

makeup system to provide positive pressure and heat the intake air. Crossdraft booths were the industry standard until the 1980s. They are still produced and are available at a lower cost than more modern booths. The major problem with a crossdraft booth is dirt. Because the air travels over the entire length of the vehicle, there is a good chance of blowing dirt from one area of the vehicle into the wet paint.

A **semi-downdraft** booth is a hybrid of cross- and downdraft booths. In this type, the air enters through the roof at one end of the booth and is exhausted through the wall at the opposite end of the booth (Figure 2–8). The air travels over the entire length of the car, causing dirt problems similar to crossdraft booth. A semi-downdraft booth may be equipped with an air make-up unit.

The intake air for a **downdraft** booth can be supplied by an air makeup unit or simply by intake filters in the roof. The exhaust air can exit through filters lo-

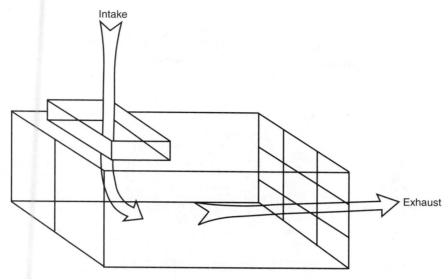

FIGURE 2-8 *Semi-downdraft booth.*

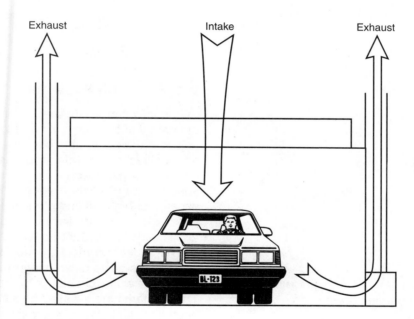

FIGURE 2-9 *Side downdraft booth, front view.*

cated in the walls near the floor—a side downdraft (Figure 2–9)—or through the floor into a pit—a full downdraft (Figure 2–10). In some applications, the entire floor is a grate over a basement. The grate has a filter and the whole floor acts as a vent. In other applications, there is a pit in the floor at the center of the booth. The vehicle is then parked over the pit. Downdraft booths minimize the amount of contaminated air flow over the vehicle because the air flow is from top to bottom, not the length of the vehicle. There is less of a chance of dirt contamination in a downdraft booth.

AIR FILTRATION SYSTEMS

The air filtration system removes many contaminants either before the air enters the booth (intake filters) or before the air exits the booth (exhaust filters). Currently there are two commonly in use (Figure 2–11):

1. Wet filtration system
2. Dry filtration system

Intake filters are dry. Exhaust filters may be wet or dry.

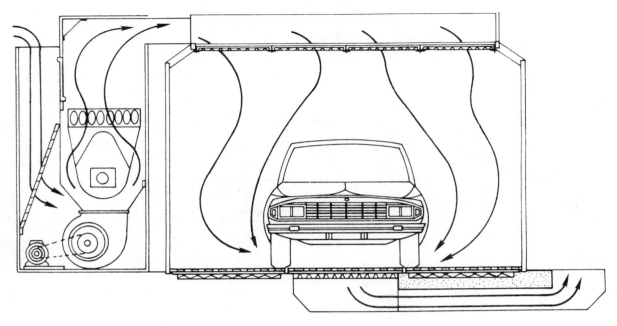

FIGURE 2–10   Full downdraft booth.

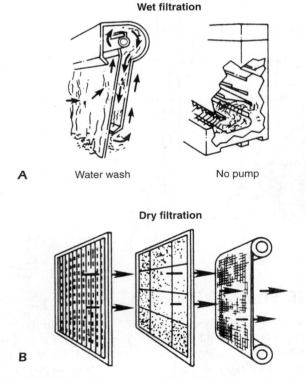

FIGURE 2–11   Spray booth filtration: (A) wet (B) dry.

Wet Filtration Systems

Although wet or wash filtration has a higher initial cost than a dry filter system, it has grown in popularity with downdraft spray booths because water filtration does an excellent job of removing paint particles from exhaust air regardless of the paint viscosity or drying speed. It can handle a variety of spray materials, is capable of high volume production, eliminates the expense and inconvenience of changing exhaust filters, and is accepted by most local fire codes.

The typical downdraft booth with a water filtration system has ducts or an open grate floor under which a layer of water circulates to carry away overspray. The contaminated water is routed through a "compounding" tank where the paint is separated out before the water recirculates through the system. Exhaust air from the booth is purified by routing it through a water curtain wash system. A continuous spray mist of water scrubs the paint particles from the air, while baffles reverse the direction of the airflow to help separate out the particles by centrifugal action (Figure 2–12).

There are wide varieties of water wash configurations available, with some offering various advantages over others. There are also "pumpless" wash sys-

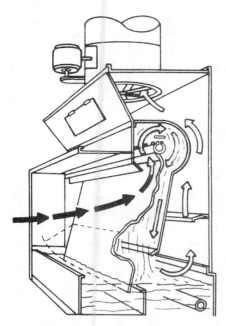

FIGURE 2–12 *The air is cleansed by a combination of water spray and centrifugal separation in a typical water wash system. (Courtesy of Binks Mfg. Co.)*

tems. Instead of using a curtain spray wash to clean the air, the air is pulled through a pan of water and a series of baffles by a high-pressure fan. Air velocity pulls the air through the water, creating a swirling mixture that washes out the paint particles. The water is treated with an anticoagulant additive so the paint particles settle to the bottom (in pump wash systems, the paint rises to the top). Pumpless wash systems are very sensitive to the water level and air pressure, so proper maintenance is very important.

The paint residue that collects in the water must be removed periodically and the water kept at the specified level. The rate at which water evaporates from the system will depend on temperature, humidity, the volume of usage, and the design of the system itself. But it is necessary to add makeup water at least once a week unless the booth has an automatic water makeup feature. Plus water additives must be placed in the water to prevent the growth of bacterial/germ growth and the creation of an unpleasant smell.

The paint residue that is separated from the water hardens. To dispose of this solid waste, check local or state regulations. It may be considered hazardous waste. One innovative approach is to recycle the residue by mixing it with undercoating and spraying it under cars. As for water disposal, there is no reason to worry about it because the water is continuously recycled. If the system has to be drained for some reason, the waste water must be disposed of in accordance with whatever local sewer restrictions apply.

The disposal of residual solid waste or waste water may be costly.

Dry Filtration Systems

These spray booth systems come in various configurations and filter media (paper, cotton, fiberglass, polyester, and so on). What is important here is the efficiency of the filtration system and how quickly the filters clog up. Dry filters work like a sieve. They mechanically filter out particles of paint and dirt by trapping the particles as air flows through the filter. Some are also coated with a tacky substance so particles will adhere to the surface of the fibers.

Most filtration systems today can remove virtually 100 percent of the particulates that are of a size large enough to cause a noticeable blemish in a paint job. Anything larger than about 14 microns (.005 inch, the smallest particle that can be seen by the naked eye) can leave a noticeable speck in the paint. Anything smaller than 14 microns is usually encapsulated in the paint and will not cause problems. Most of the filters that are used in the ceilings of downdraft booths or the doors of crossdraft booths today will stop anything larger than about 10 microns from getting through.

As a filter traps more and more particles, it becomes more dense and thus more efficient. But at the same time, it also offers increasingly greater resistance to the flow of air. Eventually the point is reached where the filter restricts airflow to such an extent that it interferes with proper airflow through the spray booth. Ideally, the filter should be changed before it reaches that point. It is something that has to be watched very closely.

The best way to judge a filter's condition is to measure its air resistance with a water column pressure differential gauge (**manometer** or magnehelic gauge). Some booths have built-in gauges, others do not. Comparing air pressure upstream of the filter to that which is downstream indicates whether or not it is time to replace filters. The amount of restriction that is considered acceptable will vary according to filter construction and media, spray booth construction and air volume. Some filters should be changed when there is as little as 0.25 inch difference in the water column side to side, while others can handle up to 1-1/2 inch of difference before they have to be replaced.

Because the amount of restriction that is considered acceptable can vary so much from one type of filter to another, it is important to check with the filter supplier before using any replacement filters that are different from those originally supplied with a spray booth. The type of filter media used will also have a significant bearing on maintenance costs, filtration efficiency, and filter longevity. Filters are not a generic product. One type might be much better suited to a particular application than another. That is why spray booth manufacturers typically put such a high emphasis on filter selection.

There are still other considerations that a paint shop operation must keep in mind when selecting a filtration system. In southern California, for example, shops over a certain size are now facing an additional air filtration expense. Neither dry nor wet filtration can remove harmful chemicals and solvents such as isocyanates from the exhaust air, so additional exhaust filtration is now being required. To date, unfortunately, the only approved methods of treating exhaust air are with an afterburner system (which is very expensive) or with activated carbon filtration. Some booth manufacturers claim that water filtration can "neutralize" isocyanates, but this has not yet been approved in California.

SPRAY BOOTH MAINTENANCE

Regardless of the type of filtration that a paint shop employs, maintenance is a prime consideration, not only from the standpoint of cost and convenience, but also because it is essential to maintain the quality of the paint job. The best air filtration system in the world will not be able to do its job if it is poorly maintained. The first task in learning how to avoid dirt is to understand where dirt comes from. Anything that is brought into the booth can bring dirt with it. Potential sources of dirt include the air, the vehicle, the painter, the equipment and supplies, and even the paint.

Incoming air is a prime source of dirt. Dirt in the air is generated by dirty filters, imbalanced air pressures, and open doors. Check the intake filters daily and change them as soon as the manometer indicates. When dust and dirt start to clog filters and restrict airflow, the velocity of air passing through the filters begins to climb. Increased velocity increases the likelihood of pulling dirt through the filters. In a booth with an air makeup system, balance the input air pressure against the exhaust air to provide slightly positive pressure in the booth. This balance constantly changes as filters load up and can change from car to car. Therefore, check and adjust it with each new job.

You may enter a positive pressure or air make-up unit equipped spray booth with the fans running. The positive pressure helps keep the dirt out. But, keep traffic in and out of the booth to an absolute minimum. Also, make sure that the collision repair facility doors are closed at all times during the painting operation. Opening and closing these doors can cause the booth air balance to fluctuate, creating turbulence and dirt inside the booth. Enter a negative pressure or no air make-up unit equipped spray booth only when the fan is off. This prevents pulling contaminated shop air inside the booth.

The booth itself can be a main contributor to dirt problems through air leaks, poor housekeeping habits, exhaust air, and floor coverings. There are proper seals for door frames, light openings, and panel seams that must be installed properly and replaced periodically. Heavy usage and temperature extremes quickly destroy these seals. Use caulking as an inexpensive gap sealant to keep dirt out of the clean airstream.

When operating the spray booth, keep the following points in mind:

1. Follow the manufacturer's recommendations for the minimum velocity needed to exhaust spray vapors properly. If that recommendation is exceeded, turbulence cancels out the screening performed by the filters. If the velocity is too low, the air will not move fast enough to remove overspray and airborne dirt before it causes defects.
2. Depending upon the volume of spraying, paint arresters are a high-consumption item requiring frequent changing. Check filter resistance daily on the manometer. When paint accumulation builds up, velocity goes down, and air movement is too slow to remove the overspray.
3. In a dry filtration system, the filters must be periodically inspected and replaced (Figure 2–13). And when they are replaced, the multistage filters designed for the booth should be used.

FIGURE 2–13 *Checking the exhaust filters on a reverse-flow crossdraft booth.*

WARNING: Clogged filters are a fire hazard, because they could catch fire, under certain conditions.

4. Be sure the water level in the wet filtration system is kept at its proper working level and that the correct water additive is used.

In order to get the best results from any type of spray booth, it is important to follow a routine good housekeeping program such as this:

- Periodically wash down the booth walls, floor, and any wall-mounted air controls to remove dust and paint particles. Some shops may require that floor and walls be wiped down after every job. Always pick up any scrap, paper, rags, and so forth.
- Periodically inspect the inside of the exhaust stack and air make-up unit. These areas may be cleaned by vacuuming or power washing.
- The booth is no place to store parts, paint, trash cans, or workbenches because dirt will accumulate on these things and will eventually land on the vehicle. Keep these items in a sealed, ventilated storage area.
- Be sure that all bodywork and most paint preparation procedures are done outside of the spray booth. Final masking may be done inside the booth. Make certain no sanding or grinding operations are performed in or near the spray booth. The dust created will go all over and ruin not only a present job but many future jobs.
- Water is most often used to contain dirt. It is cheap and effective at trapping dirt, but it can splash on the car midway through the job, or, in a heated booth, dry out before the paint job is finished. If water is sprayed on the floor to keep any stray dust down on the floor, be sure to eliminate all puddles to prevent splashes. Water can rust the walls of the spray booth, resulting in premature deterioration. Also, the high humidity from evaporating water may cause isocyanate-containing paints to cure faster than normal, resulting in problems.
- Roofing felt held to the floor with duct tape provides an inexpensive method of containing dirt. It attracts and holds lint, lasts longer than water, is not a hazardous waste, and does not deteriorate in the booth.
- Clay tiles look nice and provide an easy-to-clean smooth surface, but they are expensive and slick to walk on.
- Concrete sealant provides a smooth, easily cleaned surface that is somewhat inexpensive.

It should be noted that any slick surface treatment adds to the turbulence, while a textured surface tends to impede the turbulence or at least scrub the air.

- Strippable spray on booth coverings can be used to trap dirt. When the overspray on the covering becomes too thick, strip and recoat.

The vehicle itself is often the greatest source of dirt in the spray booth. Dirt hides in cracks and crevices, behind bumpers, and in the engine compartment. Even a thoroughly cleaned vehicle collects dirt when left in the general sanding area before being brought into the booth. When the spray gun hits this dirt at 50 psi, it kicks it out of its hiding places and deposits it into the finish. That is why a good prep job is so important.

SHOP TALK

A cotton T-shirt is perhaps the greatest source of contamination and should never be worn in the booth. Lint-free paint suits, rubber form-fitting gloves, a dirt-free head cover, and the appropriate respirator should be worn inside the booth. Remain in the booth between the application of coats rather than risk dragging dirt back into the booth. If this step is not possible, remove the protective suit inside the booth and leave it there. Upon returning, put the suit back on to contain the dirt collected on the clothing worn outside the booth. (Anyone not interested in wearing the proper attire should view the work through an observation window rather than risk exposure to hazardous vapors.)

Spray guns and cups should always be kept spotless inside and out. Do not use those dirt-collecting cloth wheel covers. Spray guns, masking paper, paint cans, tape, wheel covers, air transformers, hoses, respirators, coveralls, tack rags, and various other supplies can all collect dirt if stored in a dirty environment. All of these items should be kept in a filtered, ventilated storage/mix room. If any of these items are subject to sanding dust, they will quickly ruin a paint finish.

Unbelievable as it might seem, dirt from compressed air lines often causes blemishes in paint jobs. Air transformers, with properly cleaned and regularly drained filters, keep the air clean and dry. Oil and water separators are absolutely necessary to eliminate dirt and contamination.

A buildup of overspray can collect on the air cap and turn into a kind of fuzz. Clean the gun frequently to prevent fuzz from blowing off and ruining a paint job. Paint will set up in and on the gun. If the dried

TABLE 2–1: TROUBLESHOOTING SPRAY BOOTH PROBLEMS

Fault	Result	Dirty Job	Thin Coats	Poor Opacity	Sags	Overloading	Popping	Softness	Overspray	Uneven Application	Recoat Failure	Fire Hazard	Water Splashes
Dirty filters in a dry filtration system	Vacuum in booth (hot air drawn from oven)	C				B	A	A	C	C,D	A		
	OR												
	Not pressurized (low air movement and dirty air drawn in from preparation area)	A	A	C	B					C,D			
Torn or damaged filter in a dry filtration system	Turbulence	A							B	B,D			
	Over-pressurized				A	A	A	A	B	C,D	A		
Water level in a wet filtration system — Low	Increased extraction					A	A	A		C,D	A		A
— High	Restricted extraction		A	C	B				C	C,D			A
— Empty	Increased extraction with buildup of dry paint reservoir	A										A	
Use of incorrect water additive, or incorrect use of water additive in a wet filtration system	Blocked water jets and filters. Formation of dry powder on anti-splash panels. Corrosion of plant. Paint deposits difficult to remove	A										C	A
Rags, masking paper, old cans, and so on in booth	Dirt accumulation	A										A	
Spraying on walls of booth	Poor light reflection									C,D			
Loose deposits of dirt, dry spray, rust, and so forth on booth walls	Dirt in atmosphere	A											

A   Most likely failure to be associated with the fault
B   Likely failure
C   Failure less likely to be associated with the fault
D   Will affect color of metallics

paint flakes off, it may land on the wet paint and cause a defect. Clean the gun inside and out after every job.

Oil fan pulley and motor bearings regularly, if required, but always be sure to shut off the main fan switch of the power supply before oiling the fan.

If the spray booth is not properly maintained, it can cause the finish problems described in Table 2–1.

2.2 — PAINT CURING

Paint **cure** time is dead, nonproductive time for a collision repair facility. To keep this dead time to a minimum, high-production shops shorten cure time by heating the refinished surface. This can be done in a baking booth or drying room or with a portable infrared heater.

A baking booth is a spray booth equipped with a high-temperature air makeup unit (Figure 2–14). After the vehicle has been sprayed, the exhaust fan continues to remove contaminated air for 10 to 20 minutes. After this time during the cure cycle, the air makeup unit increases the intake air temperature to 140 degrees. The volume of the intake air is reduced to half of what was used during the spray cycle. One manufacterer's paint may be baked at 140 degrees for 30 minutes reducing the cure time from eight hours to one hour. In a high-production shop, this time savings means that more vehicles can be repaired per week.

Another method of rapid cure is to place the refinished vehicle in a separate room (Figure 2-15). The drying room is usually attached to the spray booth. After the vehicle is sprayed and purged, it is moved into the drying room. In some cases, the drying room is heated to 140 degrees by an intake air furnace. In other applications, a series of infrared heaters line the

FIGURE 2–14   The temperature control for a spray booth equipped with a high-temperature air makeup unit.

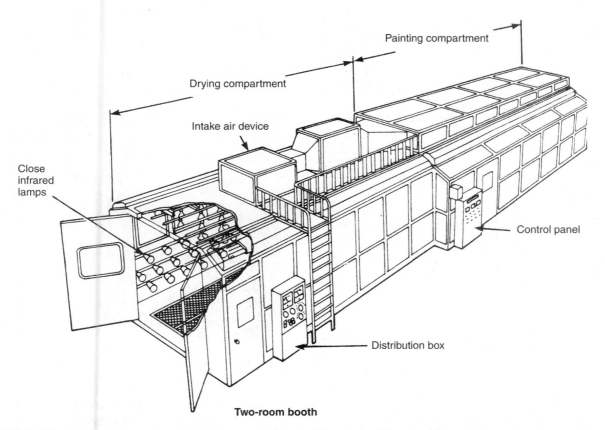

Two-room booth

FIGURE 2–15   Drying room separate from a spray booth. (Courtesy of Toyota Motor Corporation.)

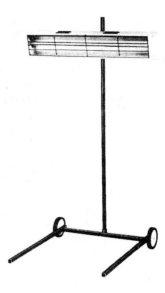

FIGURE 2–16 *Various types of infrared heaters.*

walls of the room. The drying room allows even more production than the baking booth, because as one vehicle cures in the drying room, another vehicle can be sprayed.

The last method of rapid paint cure is a portable infrared heater (Figure 2–16). These units are becoming more popular because they increase productivity on panel repairs. There are two types of infrared heaters: shortwave and medium/longwave.

Medium/longwave infrared curing works by heating the paint. The heat travels downward toward the panel. This "top-down" curing can lead to problems. If the surface hardens before the paint underneath it dries, the solvent is trapped. Trapped solvent can lead to solvent pop, shrinkage, and die back. This is not the best method.

The shortwave heaters warm the panel itself. The heat travels outward, causing "bottom-up" curing. This method avoids the problems associated with top-down curing. Shortwave infrared curing is also useful for drying water-based paint and primer.

 ## 2.3 — PREP STATIONS

Prep stations are used to perform sanding, priming, and other pre-paint operations. Usually prep stations consist of a fan or fans, filters, and ductwork. Some are surrounded by plastic curtains (Figure 2–17). Some types are much like a crossdraft booth with no intake filter; air is simply filtered and exhausted outside. This pulls sanding dust out of the work area. In

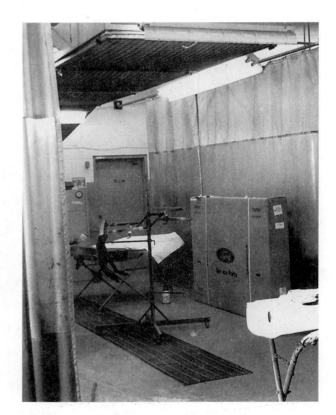

FIGURE 2–17 *Prep station.*

cold climates, this is extremely wasteful of heat. To solve this heat loss problem, some prep stations recycle the air. Dust from around the car is picked up by a stream of air pulled toward a set of filters. The cleaned air is pushed back to the car by another fan. In some cases, the air flow is semi-downdraft; in oth-

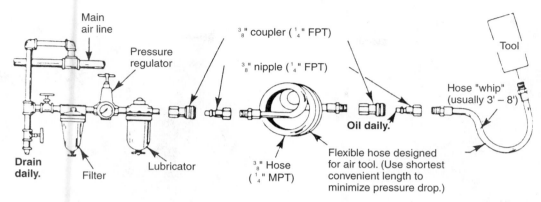

FIGURE 2-18 *Typical auto shop arrangements for air tools (but not spray guns).*

ers, it is full downdraft with an exhaust pit. Other systems are portable.

If allowed by local code and if the prep station is surrounded by a plastic curtain, the simple operations of priming and part trimming can be performed at the station. This allows the spray booth to be used for painting only, a much more efficient use of the equipment.

2.4 — PAINT MIXING ROOM

Paint mixing rooms are used to store paint code books, paint tints, mixing scales, paint gun cleaners, solvents, and paint materials. Mixing rooms can be purchased as standalone units, similar to spray booths in construction. In other cases, a separate room is constructed as part of the paint shop. In any case, the room should have power ventilation and explosion-proof lighting. If the lighting is color balanced, the room can be used for color matching.

2.5 — AIR-POWERED TOOLS

The automotive industry was one of the first industries to see the advantages of air-powered tools. Although electric grinders, polishers, and heat guns are found in some refinishing shops, the use of air tools is a great deal more common. Air tools have four major advantages over electrically powered equipment in the auto repair/paint shop:

1. **Flexibility.** Air tools run cooler and have the advantage of variable speed and torque; damage from overload or stalling is eliminated. They can fit in tight spaces.

2. **Light weight.** The air tool is lighter in weight and lends itself to a higher rate of production with less fatigue.

3. **Safety.** Air equipment reduces the danger of fire hazard in some environments where the sparking of electric power tools can be a problem. Also, often water from wet sanding is on the floor of a paint preparation area. Air tools do not have the shock hazard associated with electric tools in a wet environment.

4. **Low-cost operation and maintenance.** Due to fewer parts, air tools require fewer repairs and less preventive maintenance. Also, the original cost of air-driven tools is usually less than the equivalent electric type.

The most common causes for any air tool to malfunction are:

- Lack of proper lubrication
- Excessive air pressure or lack of it
- Excessive moisture or dirt in the air lines

The installation of an air transformer and lubricator (Figure 2-18) will greatly reduce air tool malfunctions. With these units installed, it is possible to assure clean air, proper lubrication of internal wear parts, and control of air pressure to suit different tool applications. Do not use a lubricator on an air line that serves a spray gun.

There are air-powered equivalents for nearly every electrically powered tool, from sanders to drills, grinders, and screwdrivers. Furthermore, there are some air tools with no electrical equivalents—in particular, the needle scaler, air file, air blowgun, and scraper.

PAINT SPRAY GUNS

The spray gun (Figure 2-19) is probably the most-used air-powered tool in the paint shop. A spray gun is a precision tool using compressed air to break up

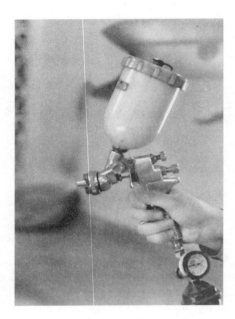

FIGURE 2–19 Gravity-feed spray gun.

A

B

FIGURE 2–21 (A) Dual action sander or DA—rough. (B) Dual action sander—finish.

FIGURE 2–20 Disc sander, also known as a mud hog.

FIGURE 2–22 Jitterbug sander.

liquid paint into small droplets. Air and paint enter the gun through separate passages and are mixed and ejected at the air nozzle to provide a controlled spray pattern. Complete details on the operation of the various types of spray guns and their use can be found in Chapter 3.

AIR SANDERS

There are two basic types of air sanders: disc and orbital (finishing). The disk on a disk sander simply spins in a circle. Most rough sanding done in automotive work is done with a disc sander (Figure 2–20). A dual-action (DA) orbital sander (Figure 2–21) oscil-

lates while it is rotating, thus creating a buffing pattern rather than the swirls and scratches often caused by the disc sander. A finish DA can be used to sand much of the vehicle. Another type of finish orbital sanders, also called a "jitterbug" sander (Figure 2–22), is designed for fine finish sanding. It is possible to use a wider variety of abrasives with finish sanders

FIGURE 2–23 Air long board or air file.

FIGURE 2–24 Disc grinder.

than with any other type of power sanding, but for the most part, the best work is done with comparatively fine grit abrasive paper. Finish sanders are also especially designed for hard-to-reach places and tight corners.

Another sander found in collision repair facilities is the long board–type sander or air file (Figure 2–23). It operates in either an orbital or straight line motion that will cover about forty square inches of working area per minute. An air file is used to rapidly level cured body filler.

AIR GRINDERS

The most commonly used portable grinder in body and paint shops is the disc-type grinder (Figure 2–24). It is operated like the disc sander. Usually a coarse disk is installed on the grinder. The grinder may be used to remove paint or metal.

CAUTION: Avoid grinding too close to the trim, bumper, or any other auto parts projections that might snag or catch the edge of the disc (Figure 2–25). If possible remove obstructions before grinding. Always stop a disc grinder while in contact with the work surface. Start any grinding operation just before the machine makes contact with the work surface.

There are, of course, several other grinders used in the paint/body shop. Following are the more common ones that a painter can use:

- **Horizontal grinder** is used for heavy-duty grinding.

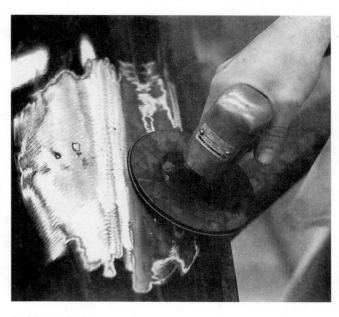

FIGURE 2–25 As the disc grinder spins, it can easily catch on an edge. Avoid edges and projections.

- **Vertical grinder** is a larger version of the disc grinder. With a sanding pad, this grinder can be converted into a disc sander. Most vertical grinders can be used with straight wheels as well as cup wheels.
- **Angle grinder** (Figure 2–26) is used primarily for smoothing, deburring, and blending welds.
- **Small wheel grinder** can be used with cone wheels, wire brushes, or collet chucks and burrs in addition to a straight grind.

FIGURE 2–26 Angle grinder, heavy duty. Removes welds and metal fast.

FIGURE 2–28   Foam pad on an air buffer.

FIGURE 2–27   Air buffer.

- **Die grinder** is used with mounted points and carbide burrs for a variety of applications such as weld cleaning, deburring, blending, and smoothing. Available in both straight and angle-head designs.

POLISHER/BUFFERS

An air-powered buffer (Figure 2–27) can be used to remove slight paint defects such as dirt. The buffer may be fitted with a wool or foam pad (Figure 2–28). The air buffer is lighter in weight than an electric buffer. Their uses are fully described in Chapter 15.

SANDBLASTERS

Sandblasting can be used to remove all paint or rust and weak metal from a vehicle. Care must be taken not to excessively damage the metal when sandblasting. Other materials besides sand may be used in some blasters. These materials may be plastic media or powders. These softer materials lessen the chance of damaging the metal. When used correctly, these softer materials can remove paint one layer at a time.

As shown in Figure 2–29, there are two basic types of sandblasters:

- Standard sandblaster
- "Captive" sandblaster

The standard sandblaster is usually operated outdoors, while the captive units can be used indoors. The indoor sandblasters have a nozzle assembly that confines the blasting action, while a vacuum in the machine sucks the abrasive and debris.

For either blaster, sand grit size number 30 is used for general-purpose applications (light rust and paint removal from metal). Aluminum oxide (grid size number 50 to 90) is used for heavy rust removal and rough-surface paint removal.

OTHER AIR TOOLS

There are several other air tools that can be found in some auto paint shops. They include:

- **Needle scaler.** Used for derusting and cleaning of metals as well as for peening welded joints (Figure 2–30A).
- **Air blowgun.** Possibly the smallest air tool in the shop, it is one of the most worthwhile. It blows away dust and dirt from any small, hard-to-reach place (Figure 2–30B).
- **Scraper.** Removing undercoating and other coverings is this accessory's function.

A
B

FIGURE 2–29　(A) Pressure sandblaster. (B) Captive sandblaster.

A
B

FIGURE 2–30　(A) Needle scaler. (B) Blow gun.

AIR TOOL MAINTENANCE

Air tools require little maintenance but could easily cause big problems if that little maintenance is not performed. For instance, moisture gathers in the air lines and is blown into the tools during use. If a tool is left with water in it, rust will form, and the tool will experience reduced efficiency and will wear out much more quickly.

To prevent this from happening, remember that air tool motors need lubrication with a good grade of air tool oil. If the air line has no line oiler or lubricator, run one or two drops of oil through the tool for each hour of use. Squirt the oil into the tool air inlet or into the hose at the nearest connection to the air supply; then run the tool. Most air tool manufacturers recommend the use of their special oil for lubricating tools.

Most air refinishing tools have a recommended air pressure (Table 2–2). If the tool is overworked, it will wear out sooner. If something goes wrong with the tool, fix it. If not, a chain reaction might occur and the other parts will require maintenance also. For example, if the gearing must be replaced and the tool is used anyway, the rotor and end plate might soon wear out as a result. A tool with worn parts will also use more air pressure. The air compressor, in turn, will then become overworked. It will put out air that is

TABLE 2–2:	AIR CONSUMPTION CHART†	
Tool	**SCFM***	**PSI***
Air blowgun	1–2.5	40–90
Airbrush	1	10–50
Car washer	8.5	40–90
Cut–off grinder	4–8	80–90
Drill* (3/8 inch)	4–6	70–90
Grinders, vertical	10–16	70–90
Needle scaler	3–4	70–90
Paint sprayer	0.7–5	10–70
Pneumatic garage door	2	90–150
Polisher	2	70–90
Riveter	4.5–5.5	70–90
Sandblast gun	2.2–4	30–90
Sandblast gun/hopper	2–6	40–90
Sander, disc	4–6	60–80
Sander, double action	6–8	60–80
Sander, finish	6–8	60–80
Sander, straight line	6–8	70–90
Screwdriver	2–6	70–90
Shears	5–8	70–90

*SCFM: Standard cubic feet per minute
PSI: Pounds per square inch
†Always check with the tool manufacturers for the actual air consumption of the tools being used. These figures are based on averages and should not be considered accurate for any particular make of tool.

not as clean or dry as it should be; this air is shot right back into the tools.

Full information on pneumatic air system operation is given in Chapter 4.

ELECTRIC-POWERED TOOLS
2.6

As mentioned earlier in this chapter, shop tools such as sanders and polishers can also be powered by electric motors. But for most paint shops, the most important electric-only tools are bench grinders, vacuum cleaners, and heat guns.

Other than these specialized electrically driven power tools, electric drills, polishers, sanders, and so on perform the same shop tasks as their air-powered counterparts.

BENCH GRINDER

This electric power tool (Figure 2–31) is generally bolted to one of the shop's workbenches. A bench grinder is classified by wheel size; six- to ten-inch wheels are the most common in auto repair shops. Three types of wheels are available with this bench tool:

- **Grinding wheel.** For a wide variety of grinding jobs from sharpening cutting tools to deburring.
- **Wire wheel brush.** Used for general cleaning and buffing, removing rust and scale, paint removal, deburring, and so forth.
- **Buffing wheel.** For general-purpose buffing, polishing, light cutting, and finish coloring operations.

When using a bench grinder (these safety rules also apply to portable air grinders), remember to:

- Always use a wheel with a rated speed equal to or greater than the grinder's.
- Always use a wheel guard.
- Inspect wheels for wear or cracks before using them. Use correct wheel for job and mount it properly.
- Always wear safety goggles or face shield and make sure the eye shields of the grinder operator are in position.
- Always wear ear protection.
- Check that the tool rest is no further than one-eighth of an inch from the grinding wheel.

FIGURE 2–31   *Bench grinder.*

FIGURE 2–32   *Standard vacuum cleaner.*

VACUUM CLEANER

A must in every body and refinishing shop is a vacuum cleaner (Figure 2–32). The vacuum cleaner can be used to remove dirt from a vehicle interior prior to paint or delivery. There are two basic types of shop vacuum cleaners: the dry pickup type and the wet/dry unit. The wet/dry unit, in the twenty- to thirty-gallon

FIGURE 2–33 Handheld vacuum cleaner.

FIGURE 2–34 Heat gun.

FIGURE 2–35 Electric buffer.

capacity, is the one found in most paint shops. For interior vehicle cleaning the portable vacuum cleaner is popular (Figure 2–33).

POWER WASHER

Power washers can be used to thoroughly clean the outside of a vehicle prior to refinishing. Some shops power wash all vehicles before they enter the paint shop. Another common practice is to power wash a vehicle after all sanding operations have been completed. Power washers can also be used in exterior car preparation, engine cleaning, undercarriage cleaning, shop degreasing and cleaning, and snow and salt removal from vehicles.

HEAT GUN

Heat guns (Figure 2–34) have a number of uses in the auto paint shop. They can be used to remove woodgrain transfers decals and glued on name-

plates. They are also helpful in stretching vinyl rooftop corners.

ELECTRIC BUFFER

An electric buffer (Figure 2–35) is used for the same purpose as an air buffer—to remove minor paint defects. An electric buffer is heavier than an air buffer, but electric buffers are usually more powerful. Some electric buffers have a variable speed setting.

ELECTRIC POWER TOOL SAFETY

To protect the operator from electric shock, most power tools are built with an external grounding system. That is, there is a wire that runs from the motor housing, through the power cord, to a third prong on the power plug. When this third prong is connected to a grounded, three-hole electrical outlet, the grounding wire will carry any current that leaks past the electrical insulation of the tool away from the operator and into the ground of the shop's wiring. In most modern electrical systems, the three-prong plug fits into a three-prong, grounded receptacle. If the tool is operated at less than 150 volts, it has a plug like that shown in Figure 2–36A. If it is for use on 150 to 250 volts, it has a plug like that shown in Figure 2–36B. In either type, the green (or green and yellow) conductor in the tool cord is the grounding wire. Never connect the grounding wire to a live terminal.

Some of the new electric power tools are self-insulated and do not require grounding. These tools have only two prongs, since they have a nonconducting housing. In shop operations, never use a three-prong adapter plug.

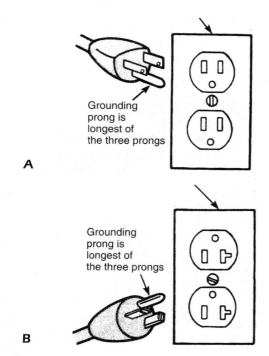

A

Grounding
prong is
longest of
the three prongs

B

Grounding
prong is
longest of
the three prongs

FIGURE 2–36 (A) Approved type three-prong grounding plug and outlet box for 115 volts. (B) Approved type three-prong grounding plug and outlet box for 230 volts.

Extension Cords

If an extension cord is used, it should be kept as short as possible. Very long or undersized cords will reduce operating voltage and thus reduce operating efficiency, possibly causing motor damage. Actually, an extension cord should be used only as a last resort. When an extension cord must be employed, the following wire gauge sizes are recommended for different lengths:

Length	115 Volts	230 Volts
Less than 25 feet	12 gauge	14 gauge
25 to 50 feet	10 gauge	12 gauge
50 to 100 feet	8 gauge	10 gauge

The smaller the gauge number, the heavier the cord. These are recommended minimum wire sizes.

Tools with three-prong, grounded plugs must only be used with three-wire grounded extension cords connected to properly grounded, three-wire receptacles (Figure 2–37).

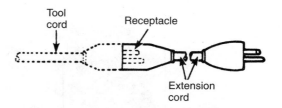

Tool cord

Receptacle

Extension cord

FIGURE 2–37 Typical three-wire extension cord.

SHOP TALK

Here are some safety tips to keep in mind when using extension cords:

- *Always plug the cord of the tool into the extension cord before the extension cord is inserted into a convenience outlet. Always unplug the extension cord from the receptacle before the cord of the tool is unplugged from the extension cord.*
- *Extension cords should be long enough to make connections without being pulled taut, creating unnecessary strain and wear.*
- *Be sure that the extension cord does not come in contact with sharp objects. The cords should not be allowed to kink, nor should they be exposed to hot surfaces or dipped in or splattered with oil, grease, or chemicals.*
- *Before using a cord, inspect it for loose or exposed wires and damaged insulation. If a cord is damaged, it must be replaced. This advice also applies to the tool's power cord.*
- *Extension cords should be checked frequently while in use to detect unusual heating. Any cable that feels more than comfortably warm to a bare hand placed outside the insulation should be checked immediately for overloading.*
- *See that the extension cord is in a position to prevent tripping or stumbling.*
- *To prevent the accidental separation of a compressor cord from an extension cord during operation, make a knot as shown in Figure 2–38A, or use a cord connector as shown in Figure 2–38B.*
- *Do not use extension cords inside the paint room.*

In recent years a few cordless tools—drills and sanders (Figure 2–39)—have made their way into the body/paint shop. These tools require no air hose or electric cord, but they require recharging.

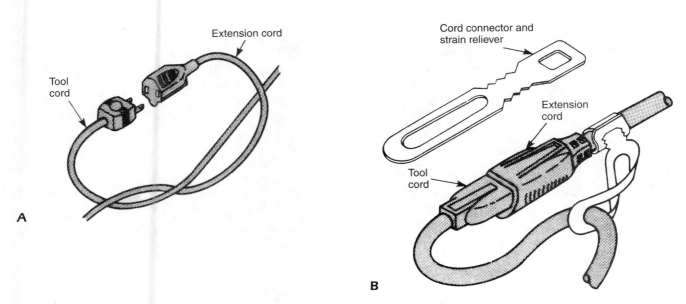

FIGURE 2–38   (A) Knot will prevent the extension cord from accidentally pulling apart from the tool cord during operation; (B) cord connector will serve the same purpose effectively.

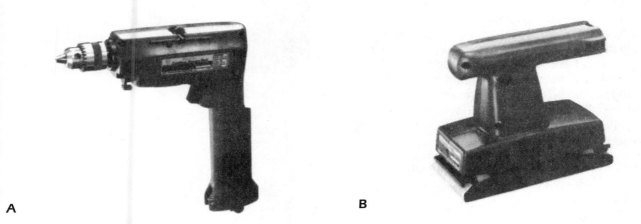

FIGURE 2–39   Typical cordless tools: (A) drill and (B) sander.

OTHER PAINT SHOP EQUIPMENT AND TOOLS

There are several pieces of paint shop equipment that can help the refinishing technician to perform paint jobs better. These items include:

- **Wet-sanding stand.** A wet-sanding stand (Figure 2–40) is used for wet sanding individual components or small parts. These cabinets are made by individual paint shops with the size and installation location depending on shop requirements and conditions.

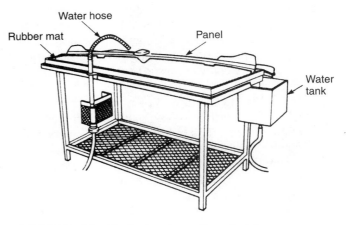

FIGURE 2–40   Typical wet-sanding stand.

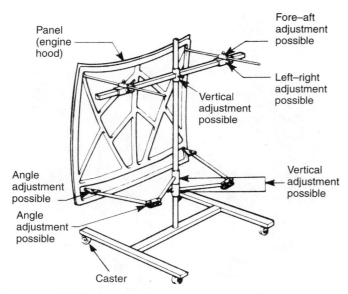

FIGURE 2–41 *Typical paint shop hanger.*

FIGURE 2–42 *Paint shaker.*

- **Paint hanger.** Paint hangers are used to suspend or secure individual components or small parts for spray painting. As with the wet-sanding stands, these are made by the individual shop in accordance with the shape of the item to be painted, the quantity required, and so on. Paint hangers keep the panel from dropping during painting. They must be made of a material that will withstand heat during paint drying. An example is shown in Figure 2–41.
- **Panel drying ovens.** These are small ovens used to dry test pieces. There are various types, from a very simple kind using infrared lamps to more complicated kinds with an electric heater, vent fan, and a timer for controlling the temperature and drying time.
- **Paint shakers.** For a good refinishing job, it is very important that the paint be thoroughly mixed or agitated. In fact, with **metallic** paint topcoats it is essential. These paints contain metallic particles that are heavier than the paint itself and quickly settle to the bottom of the container. For this reason, metallic paint, as well as most other types, needs a proper mixing job. The quickest method of achieving this is with a paint shaker (Figure 2–42).
- **Paint Mixing System.** A paint mixing system allows a refinisher to read paint formulas and combine color tints by weight to make a specified color. Some parts of the system are blade agitators (Figure 2–43) to stir tints, a mixing bank (Figure 2–44) to store tints and power the agitators. A computerized scale (Figure 2–45) is used to weigh tints while mixing.

FIGURE 2–43 *Paint blade agitator used in mixing system tints.*

FIGURE 2–44   Paint mixing system.

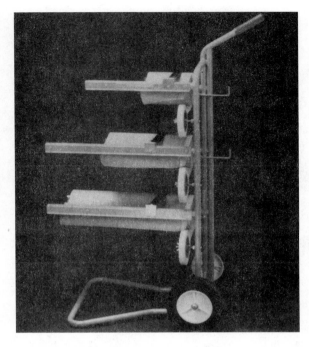

FIGURE 2–46   Masking paper and tape dispensing machine.

FIGURE 2–45   Paint scale.

- **Masking paper dispenser.** A masking paper dispenser (Figure 2–46) allows dispensing of both masking paper and masking tape at the same time; as the paper is pulled, masking tape automatically adheres to the paper edge. Two or three sizes of roll paper can be set in the dispenser to help upgrade work efficiency.
- **Metal Paint Cabinet.** Paint cabinets are used for storage and stock control of paint, reducer, undercoat, and hardener. Check local fire code for construction requirements. The cabinet should be fireproof and vented.
- **Surform file.** Body filler should be made level to the adjacent panel with a surform file (Figure 2–47). Commonly referred to as a cheese grater, the surform file is used to shape body filler while it is semi-hard. Shaping the filler before it hardens shortens the waiting period while the filler cures and reduces the sanding effort later in the repair process.
- **Speed file.** Once the body filler has hardened, the repair can be shaped and leveled with a speed file (Figure 2–48). The speed file is a rigid

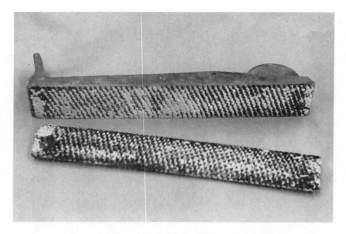

FIGURE 2–47 Surform file or cheese grater.

FIGURE 2–49 A filler spreader.

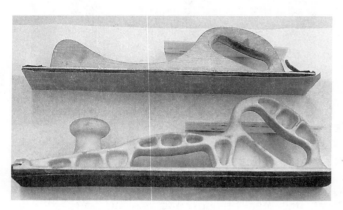

FIGURE 2–48 Speed files (top) wooden (bottom) plastic.

wooden holder about 17 inches long and 2-3/4 inches wide. Also called a flatboy, the speed file allows a repair area to be sanded quickly with long, level strokes. This eliminates waves and uneven areas.

- **Spreaders and squeegees.** Spreaders and squeegees are two important tools used in auto body resurfacing. Spreaders are used to apply body filler. Spreaders (Figure 2–49) are made of rigid plastic and are available in various sizes. Be sure to use one that is large enough to apply plastic filler over the repair area before the filler begins to set up.
- **Spray gun cleaner.** A spray gun needs to be cleaned of all dried paint after each use. Catalyzed paint (paint that has a hardener added to it) left in a spray gun will cure and possibly ruin the gun. In the past, painters simply filled the cup with thinner, hooked up the air, and pulled the trigger. This creates a fire hazard as well as air pollution. These problems can be solved with a spray gun cleaner. One type of gun cleaner is an airtight box with a sump for thinner, a pump to

FIGURE 2–50 Paint gun cleaner.

move the thinner, and nozzles to spray the thinner over the gun. The painter places the dirty spray gun into the cleaner, closes the lid, and the gun cleaner does the work. Thinner is pumped through the paint passages in the gun. The outside of the gun is sprayed and the inside of the cup is flushed. The gun cleaner can also be used on paint mixing cans, paint mixing sticks, and other paint-contaminated equipment (Figure 2–50).

FIGURE 2–51 *Various grits and sizes of sandpaper.*

BASIC PAINT SHOP MATERIALS

Paint shop materials differ from refinishing equipment in one key way: Paint shop refinishing materials are expendable. They are used up in the day-to-day operation, whereas equipment is used over and over.

Popular paint shop materials include:

- Abrasive paper
- Clean cloths or paper towels
- Tack rags
- Paint paddles
- Strainers
- Containers
- Masking paper and tape

ABRASIVE PAPER OR SANDPAPER

The rough side of paper is called the *grit* side. Grit sizes vary (Figure 2–51) from coarse to micro fine and are ordered by number. The lower the number, the coarser the grit.

CLEAN CLOTHS OR PAPER TOWELS

It is important that the areas to be painted are clean. Most paint shops provide clean cloths or special disposable paper wipers. Whichever is used, these simple tips will help:

- Use clean, dry cloths folded into a pad.
- When using a cleaning solvent, make sure to pour enough onto the pad being used to thoroughly wet the surface to be cleaned.
- Do not wait for the solvent to dry. Wipe it dry with a second clean cloth.
- Refold the cloth often to provide a clean section.
- Change cloths often.
- Once an area is clean, do not touch it with the hands, as it might affect adhesion.

TACK RAGS OR CLOTHS

These are specially treated sticky cloths (varnish-coated cheese cloth) that are used in the makeready operation to wipe the surface clean just before the paint is applied (Figure 2–52). They should be used on the area to be painted to remove all traces of lint and dirt specks. If a painter were to simply blow off the area with an air nozzle or will use an old rag to wipe off the area, he or she will leave minute impurities on the surface that will detrimentally affect the adhesion and performance of the paint. A tack cloth will pick up fine particles that are invisible to the naked eye. Tack rags should be stored in an airtight container to conserve their tackiness.

PAINT PADDLES

Made of either wood, metal, or plastic, paint paddles are used to stir the paint material. If the paddle is wood, it is recommended that the end be tapered to a sharp edge like a chisel. This will make it easier to dislodge the pigment and metallic flakes from the bottom of the can.

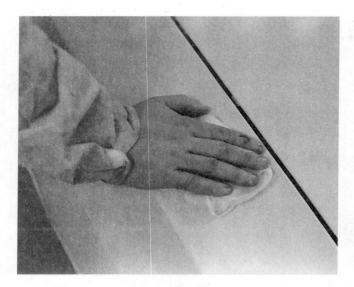

FIGURE 2-52 Painter using a tack rag to remove dust from a surface prior to painting.

FIGURE 2-53 Paint strainer used to filter out particles from mixed paint.

STRAINERS

Consisting of a cardboard funnel with cotton mesh, a strainer is used when pouring thinned topcoat and other materials into a spray cup (Figure 2–53). This is done to make sure it is free of any dirt or foreign material.

CONTAINERS

Lids on round paint product containers of a gallon or less—called *friction lids*—should be carefully opened with a proper opener. After pouring off whatever amount of material is needed from a round can, the lip should be wiped and the lid replaced tightly to form a good seal . A pouring spout made of masking tape will keep liquid from collecting in the rim, while a rubber mallet is recommended for tapping the lid around the edge. Proper resealing of the can will keep air out and minimize the formation of the film on the top called "skinning." It will also prevent the loss of solvents. Screw-top cans should be carefully wiped and closed tightly for the same reasons.

SHOP TALK _____

Keep the pouring spouts of the stirring heads clean to avoid a buildup of paint residues around the spout that can ultimately affect the accuracy of pouring.

Wiping the spout after every pour is the simplest method. Alternatively, an application of masking tape around the spout that can be replaced at regular intervals to remove solidified residues is fairly effective.

SHOP TALK _____

Store clean empty containers upside down to prevent the entry of dirt or other forms of contamination.

Plastic measuring cups are used for mixing paint. Generally, their sizes range from one quart to five quarts, and they are made of easy-to-use and easy-to-clean plastic. Disposable plastic mixing cups are also available.

MASKING PAPER AND TAPE

Masking paper is used to cover surrounding areas not to be painted so that paint overspray does not settle there (Figure 2–54). It is necessary that the masking paper be capable of preventing solvent in the paint mist from seeping through to the surface of the vehicle. There are different sizes and grades of masking paper available. Do not use newspapers for this purpose because thin fibers from the paper come off and adhere to the freshly painted surface, resulting in dirt.

FIGURE 2–54   A masked vehicle.

Also, solvent will seep through newspaper or even transfer newsprint onto the area covered.

Masking tape is used to stick the masking paper to the areas to be covered, or it can be used by itself. Masking tape is made of different types of materials, such as paper, cloth, and vinyl, so that adhesion performance is assured regardless of the reason or weather. The adhesive performance of masking tape does not change when heat is applied, and it will not leave traces of adhesive when removed. Also, it is easy to cut or tear off.

There are several types of paper masking tape, depending on what it is used for, but they can be roughly classified into general masking tape used for air drying and heat-resistant tape used for baking enamel. The proper tape for the job must always be used.

REVIEW QUESTIONS

1. Technician A says in a crossdraft booth, the air flow comes in from the ceiling and out of the end wall. Technician B says that in a semi-downdraft booth, the air flow comes in from one end and out the other end. Who is right?
   a. technician A
   b. technician B
   c. both technician A and technician B
   d. neither technician A nor technician B

2. Which booth design can have air makeup?
   a. reverse flow crossdraft
   b. regular flow crossdraft
   c. downdraft
   d. a and c only

3. Technician A says that heated air makeup is needed in cold climates. Technician B says that a negative pressure booth has an intake fan only. Who is right?
   a. technician A
   b. technician B
   c. both technician A and technician B
   d. neither technician A nor technician B

4. A pit in the floor is a feature in what type of booth?
   a. crossdraft
   b. semi-downdraft
   c. downdraft

5. Technician A says that dirt on the painter contributes to dirt in the paint finish. Technician B says that dirt on the vehicle contributes to dirt in the paint. Who is right?
   a. technician A
   b. technician B
   c. both technician A and technician B
   d. neither technician A nor technician B

6. Vehicles can be painted in a prep station.
   a. True
   b. False

7. Technician A says that an air file is also called a board sander. Technician B says that board sanders can operate either straight line or orbital. Who is right?
   a. technician A
   b. technician B
   c. both technician A and technician B
   d. neither technician A nor technician B

8. Air tools should be oiled before use.
   a. True
   b. False

9. Technician A uses a DA to sand paint. Technician B uses a jitterbug to sand paint. Who is right?
   a. technician A
   b. technician B
   c. both technician A and technician B
   d. neither technician A nor technician B

10. _____ are used to remove dirt and lint from a surface prior to painting.

Chapter 3

Spray Gun and Its Use

Objectives

After reading this chapter, you will be able to:
- Identify the spray painting equipment used in auto refinishing.
- Explain how a spray gun works.
- Identify the basic techniques of good spray painting and recognize variables that influence the quality of the spray finish.
- Adjust the spraying equipment to test and develop a good spray pattern.
- Implement the stroke technique procedure for coat application, and recognize common errors made by apprentice refinishers.
- Identify the various types of spray coats.
- Clean and properly care for a spray gun.
- Identify situations for which airless spray systems or airbrushes are recommended.

Key Term List

atomization
siphon feed
gravity feed
pressure feed
basecoat
primer
surfacer
sealer
HVLP
overlap

The spray gun (Figure 3–1) is the key component in a refinishing system. It is a precision engineered and manufactured tool, and each type and size available is specifically designed to perform a certain number of tasks. Even though all spray guns have many parts and components in common, each gun type or size is suited for only a certain defined range of jobs. As in most other areas of refinishing work, having the right tool for the job goes a long way toward getting a professional job done right in minimum time.

3.1 — SPRAY GUN PARTS

The principal parts of a typical spray gun are illustrated in Figure 3–2. Some guns are equipped with a removable spray head unit containing the air cap, fluid tip, and fluid needle. Among these parts are:

- Air cap or nozzle
- Fluid tip or nozzle
- Fluid needle valve
- Trigger

FIGURE 3–1 Spray gun in use.

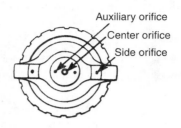

FIGURE 3–3 Parts of an air cap.

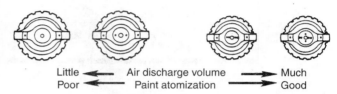

| Little | ← | Air discharge volume | → | Much |
| Poor | ← | Paint atomization | → | Good |

FIGURE 3–4 Number of holes and gun performance.

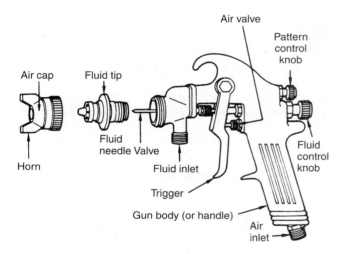

FIGURE 3–2 Parts of a siphon-feed spray gun.

- Fluid control (or spreader) knob
- Air valve
- Pattern (or fan adjustment) control knob
- Gun body (or handle)

The air cap directs the compressed air into the paint stream as it leaves the fluid tip to atomize it and form the spray pattern. There are three types of orifices (holes) (Figure 3–3): the center orifice, the side orifices or ports, and the auxiliary orifices. Each of the orifices has a different function. The center orifice located at the fluid tip creates a vacuum for the discharge of the paint. The side orifices determine the spray pattern by means of air pressure, and the auxiliary orifices promote atomization of the paint. Figure 3–4 illustrates the relationship between the auxiliary orifices and the gun's performance. Large orifices increase the ability to atomize heavy bodied or high viscosity materials such as high build surfacer. Fewer or smaller orifices usually require less air, produce smaller spray patterns, and deliver less material to conveniently paint smaller objects or apply coatings at lower speeds.

Air also flows through the two side orifices in horns of the air cap. This flow forms the shape of the spray pattern. When the *pattern or fan control valve* is closed, the spray pattern is round. As the valve is opened, the spray becomes more oblong in shape.

The *fluid needle valve* and the *fluid tip* direct the flow of material from the gun into the airstream. The fluid tip forms an internal seat for the fluid needle that shuts off the flow of material. The amount of material that actually leaves the front of the gun depends on the size of the fluid tip opening provided when the needle is unseated from the tip. Fluid tips are available in a variety of sizes to properly handle materials of various types and viscosities and to pass the required volume of material to the cap for different speeds of application. In many paint guns, the fluid

FIGURE 3–5 *This picture shows the atomization or breakup of paint into droplets.*

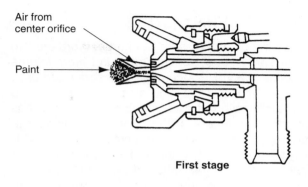

First stage

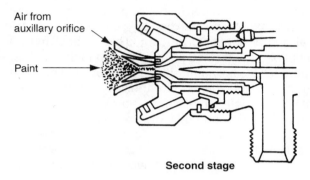

Second stage

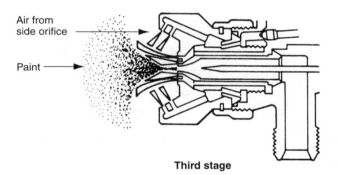

Third stage

FIGURE 3–6 *The three stages of atomization.*

tip, fluid needle, and air cap are a matched set. Charts specifying the proper fluid tip, fluid needle, and air cap for different types of paint are available from spray gun manufacturers and paint companies. The *fluid control knob (valve)* changes the distance the fluid needle valve moves away from its seat in the fluid tip when the *trigger* is pulled.

The *air valve,* like the fluid valve, is opened by the trigger. When the trigger is pulled partway, the air valve opens. When it is pulled a little farther, the fluid valve opens.

3.2 ATOMIZATION

A spray gun can be defined as a tool that turns a liquid into tiny droplets by means of air pressure. This process is called **atomization** (Figure 3–5). A thorough understanding of atomization is the key to using a spray gun correctly. Atomization breaks paint into a spray of tiny, uniform droplets. When properly applied to the auto's surface, these tiny droplets will flow together to create an even film thickness with a mirror-like gloss.

Atomization takes place in three basic stages (Figure 3–6):

- In the first stage, the paint siphoned from the fluid tip is immediately surrounded by air streaming from the center orifice. This turbulence begins the breakup of the paint.
- The second stage of atomization occurs when the paint stream is hit with jets of air from the auxiliary orifices. These air jets keep the paint stream from getting out of control and aid the breakup of the paint.
- In the third phase of atomization, the paint is struck by jets of air from the side orifices in the air cap horns. These air streams hit the paint from opposite sides, causing the paint to form into a fan-shaped spray.

TYPES OF SPRAY GUNS

In classifying by air pressure discharge, there are two types of spray guns: conventional and high volume low pressure (HVLP). A conventional spray gun uses a small volume of air at high air pressure (up to 65 psi) for atomization. If a conventional spray gun has 50 psi of air pressure at the inlet, it will discharge about 50 psi of air pressure for atomization. An HVLP spray gun uses a high volume of air at low pressure (10 psi or less) for atomization. If an HVLP spray gun has 50 psi of air pressure at the inlet, it will, depending on type, discharge 10 psi or less of air pressure for atomization. The transfer efficiency of a conventional gun is about 30 percent. This means that 30 percent of the atomized paint is applied to the vehicle; 70 percent of the paint sprayed is not applied. The pain that is not applied is either overspray—blown past the vehicle or offspray—blown back at the painter. The over- and offspray ends up on the masking paper, on the floor, and in the spray booth exhaust filters. Not only is this high pressure spraying wasteful of materials it causes excessive air pollution in the form of VOCs. An HVLP spray gun can prevent some waste and air pollution. Because an HVLP spray gun uses low air pressure to atomize paint, there is less chance of over- and offspray. More paint is applied to the vehicle; less is wasted. The transfer efficiency of an HVLP spray gun is about 65 percent.

Spray guns can also be classified by how the paint is supplied. Both conventional and HVLP spray guns can use any of the followinng methods to feed paint into the spray gun (see Figures 3–7 and 3–8, Table 3–1):

- Siphon feed. Paint cup below the gun; air flow siphons the paint out of the cup and through the gun.
- Pressure feed. The paint cup is pressurized and paint is pushed through the gun.
- Gravity feed. Paint cup is above the gun and paint flows by gravity through the gun.

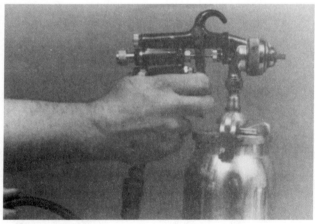

A

B

FIGURE 3–8 (A) Siphon-feed, (B) pressure-feed spray guns.

FIGURE 3–7 Paint feed methods.

The siphon-feed spray gun is used in auto refinishing shops for all types of work (spot, panel, and overall). The pressure-feed spray gun is mainly used for overall painting of vehicles (including trucks and vans), for spraying some heavier refinishing materials that are too heavy to be siphoned from a container, or where volume painting is required. Gravity-feed spray guns are becoming quite popular for all types of work.

C

FIGURE 3–8 continued *(C) gravity-feed spray guns.*

3.3 CONVENTIONAL SPRAY GUNS

SIPHON

In a siphon-feed–type spray gun the paint material is held in a one-quart cup attached to the gun. When the spray gun trigger is partially depressed, the air valve opens and air rushes through the gun. As the air passes through the openings in the air cap, a partial vacuum is created at the fluid tip (Figure 3–9A). Further squeezing of the trigger withdraws the fluid needle from the fluid tip. The vacuum sucks paint from the cup, up the fluid inlet, and out through the open fluid tip. Atmospheric air enters the cup through the air hole as the siphoned paint leaves (Figure 3–9B). The inlet air vent holes in the cup lid *must* be open for the spray gun to operate properly.

Some siphon gun cups also have an agitator system (Figure 3–10). These cups provide constant mixing of paint; they even keep large metal flakes in total suspension and complete dispersion.

TABLE 3–1: TYPES OF AIR SPRAY GUNS

Type	Paint Feed Method	Advantages	Disadvantages
Siphon-feed type	Paint container is installed below the spray nozzle and paint is supplied by suction force alone.	Stable gun operation. Easy to refill container or make color changes.	Difficult to spray on horizontal surfaces, and some variations occur in discharge volume due to variations in viscosity. Has a larger paint container than gravity-feed type, but this causes quicker painter fatigue. Cannot spray upside down.
Pressure type	Paint is pressurized by a compressed air tank or pump.	Large surfaces can be painted without stopping to refill container. A paint with a high viscosity can also be used. Can spray when held upside down.	Not suitable for small area painting. Color changes and gun cleaning take time.
Gravity-feed type	As the paint cup is installed above the spray nozzle, paint is supplied by gravity and a suction force at the nozzle tip.	Stable gun operation, all of the paint in the cup can be used.	Cup capacity is small, so not useful for painting larger surfaces; cannot spray upside down.

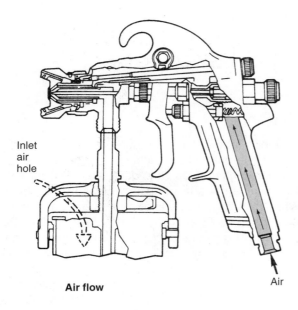

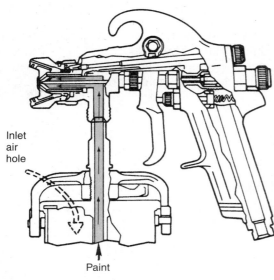

Air flow

Inlet
air
hole

Air

Inlet
air
hole

Paint

Paint flow

FIGURE 3–9   *Air and paint flow in a siphon-feed spray gun.*

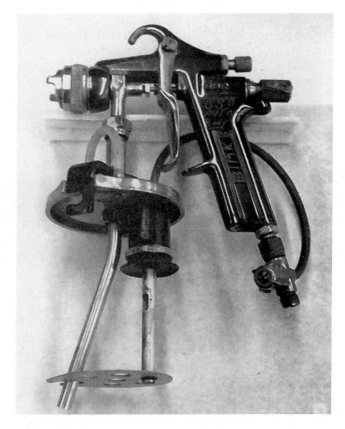

FIGURE 3–10   *Agitator system. Air is used to operate the agitator.*

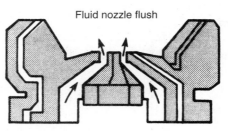

Fluid nozzle protrudes

Siphon air nozzle

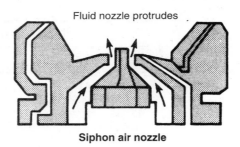

Fluid nozzle flush

Pressure-feed air nozzle

FIGURE 3–11   *The difference between siphon and pressure-feed gun air nozzles.*

PRESSURE FEED

In the design of an air pressure-feed gun, the fluid tip is flush with the face of the air cap (Figure 3–11) and no vacuum is created. The paint is forced to the air cap by pressure kept on the material in the fluid cup or tank. When equipped with a separate paint cup or tank, the pressure-feed gun is lightweight. The painter holds only the gun; the paint cup or tank can be left on the floor. The gun can be held upside down to

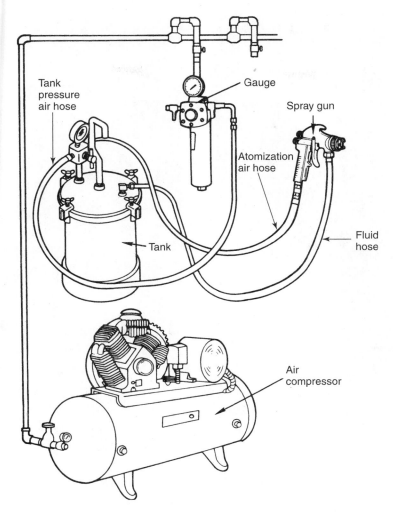

FIGURE 3–12 *Hookup for pressure tank.*

spray the underside of a vehicle. Figure 3–12 illustrates how the regulated pressure tank is hooked up for spraying:

Paint pressure tanks are available in sizes from two quarts to ten gallons. They are available in dual, single, or nonregulated models. Dual air regulators control both material and atomization air pressure; single models regulate material pressure only. Some tanks have an agitation paddle system to keep the pigments and solids thoroughly mixed at all times, assuring color uniformity.

GRAVITY FEED

Gravity-feed paint guns are easy to recognize because the paint cup is above the gun (Figure 3–13). The paint flows into the paint passage in the gun when the trigger is pulled. Although gravity-feed guns look awkward, they are well-balanced and easy to use. Unlike siphon feed guns, gravity-feed guns allow

FIGURE 3–13 *Gravity-feed spray gun.*

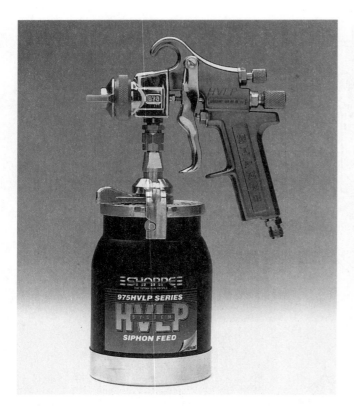

FIGURE 3–14 *A siphon-feed HVLP spray gun. (Courtesy of Sharp Manufacturing Company.)*

FIGURE 3–15 *A pressure-feed HVLP spray gun.*

all paint in the cup to be used. However, like a siphon feed spray gun, they do not work when held upside-down. Gravity-feed gun cup volume is usually about 600 ml. This means that a gravity-feed gun must be refilled quite often when one is painting a large area. The vent hole in the lid must be kept open at all times.

 3.4 — **HVLP SPRAY GUNS**

- Siphon. Siphon-feed HVLP spray guns look quite similar to conventional siphon-feed guns (Figure 3–14). Like all HVLP spray guns, this gun requires a large volume of air for atomization.
- Pressure Feed. Pressure-feed HVLP paint guns have a pressurized cup that is attached to the gun (Figure 3–15). This setup looks much different from a conventional pressure-feed gun and more like a siphon feed spray gun. When using this gun, be sure to release the pressure before opening the cap.
- Gravity Feed. Gravity-feed HVLP spray guns are similar in appearance to conventional gravity-feed spray guns (Figure 3–16).

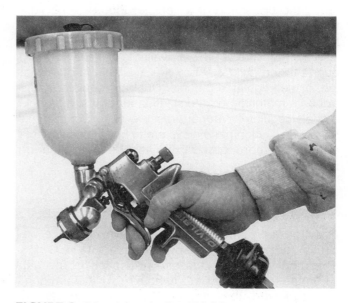

FIGURE 3–16 *A gravity-feed HVLP spray gun.*

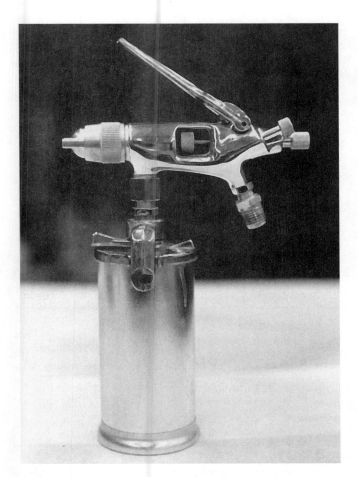

FIGURE 3–17   *A siphon-feed detail gun.*

- Detail. Detail paint guns are used to spray a small amount of paint on a small area. They can be conventional or HVLP, gravity or siphon feed (Figure 3–17). These small spray guns are very useful for painting a tight area, such as an interior or engine compartment, where a full-sized paint gun will not fit.

3.5 SPRAYING TECHNIQUES

Spraying a vehicle is a skilled job and calls for considerably more experience and knowledge than just holding down the trigger and hoping that the gun will put the paint where it is supposed to be and in the

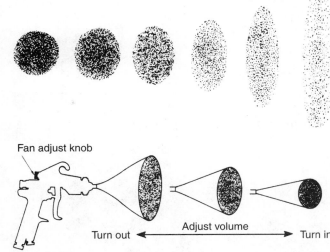

FIGURE 3–18   *Fan adjustments.*

right amount. There are several variables contributing to the quality of the finish, including spray gun adjustments and the skill of the painter.

ADJUSTING THE SPRAY GUN

A good spray pattern depends on the proper mixture of air and paint droplets, much like a fine-tuned engine depends on the proper mixture of air and gasoline. The paint must be diluted to the proper viscosity before it can be sprayed. Viscosity is a measurement of the consistency of paint or a balance of paint and solvent. The sprayed material should go on smoothly in a medium to wet coat without sagging or running. There are three basic adjustments, which under normal conditions will give the proper spray pattern, degree of wetness, and air pressure for siphon and gravity guns. Because fan size, fluid, and air pressure are all interrelated, changing one will change the others. Fine tuning of each will be needed to set the adjustments.

1. First set the fan. If you are painting a complete vehicle, open the fan up by turning out the fan adjustment knob (Figure 3–18). If you are painting a smaller area, adjust the fan to decrease wasteful overspray. If you are painting a small area such as the inside of a door, keep the overspray to a minimum by adjusting the fan to a narrow pattern.

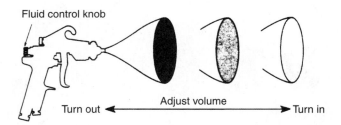

FIGURE 3–19 *Fluid adjustments.*

FIGURE 3–20 *Set air pressure with the trigger pulled slightly to release the air only.*

2. Next, set the fluid control. This determines the amount of paint that mixes with the fan (Figure 3–19). If the fan is set at maximum, the fluid should be at maximum for a wet pattern. If you desire a drier spray, you can turn the fluid knob so that less paint is mixed with the air. If you narrow the fan, you must decrease the fluid to avoid excessive material and runs. The fluid control knob can be adjusted between coats of paint to change from a medium wet coat to a wet coat.

3. Lastly, set the air pressure. The best method is to have the air pressure gauge at the gun (Figure 3–20). Pull the trigger and set the air control knob to the air pressure specified by the paint manufacturer. Different types of paint are sprayed at different air pressures, so always check the paint can or refinish guide.

TABLE 3–2: RECOMMENDED AIR PRESSURE RANGES (PSI)

| | Conventional Gun | | HVLP Gun | |
	Siphon	Gravity	Siphon	Gravity
Undercoats				
Self-etching primer	30–45	25–35	6–8	6–8
Epoxy primer	30–40	30–35	6–8	6–8
Urethane surfacer	30–40	25–35	8–10	8–10
Sealer	30–45	25–35	6–8	6–8
Adhesion promoter	30–35	25–30	6–8	6–8
Topcoats				
Single-stage enamel	45–55	35–45	8–10	8–10
Single-stage urethane	45–55	35–45	8–10	8–10
Basecoat	35–45	35–45	6–8	6–8
Clear coat	35–45	30–40	6–8	6–8

SHOP TALK

In most cases the fluid control knob is left set in the full open position. The full open position is attained when two or three threads are showing on the adjusting valve.

The optimum spraying pressure is the lowest needed to obtain proper atomization, fluid flow rate, and fan width. A pressure that is too high results in excessive paint loss through overspray (and therefore high usage) and poor flow, due to high solvent evaporation before the paint reaches the surface being sprayed.

A pressure that is too low gives a paint film poor drying characteristics, due to high solvent retention, and makes it prone to matting and sagging. The recommended pounds of air pressure vary with the kind of material to be sprayed. The typical ranges are given in Table 3–2. Conventional spray gun air pressure is measured at the inlet; HVLP spray gun pressure is measured at the air cap.

BALANCING PRESSURE-FEED GUN SYSTEM

To balance the pressure tank (Figure 3–21) or cup for spraying, the procedure is as follows:

1. After paint is poured into the container, open the pattern control knob for maximum pattern size and open the fluid control knob until the first thread is visible.

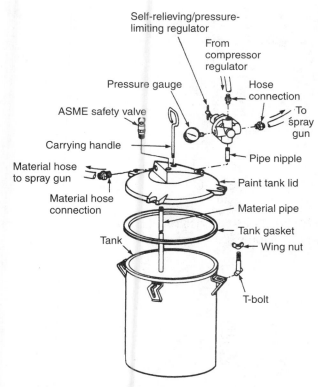

FIGURE 3–21   *Parts of a pressure pot.*

FIGURE 3–22   *Spraying a test pattern on masking paper taped to a spray booth wall.*

2. Shut off the atomization air to the gun. Set the fluid flow rate by adjusting the air pressure in the paint container. Use about 6 psi for a remote cup and about 15 psi for a two-gallon or larger container.

3. Remove the air cap, aim the gun into a clean container, and pull the trigger for ten seconds. Measure the amount of material that flowed in that time and multiply by 6 (or thirty seconds and multiply by 2). This is the fluid flow rate in ounces per minute. For standard refinishing it should be about fourteen to sixteen ounces per minute. If the flow rate is less than this, increase the air pressure in the container and repeat. If it is faster than this, decrease the pressure slightly. When the flow rate is correct, reinstall the air cap.

4. Turn on the atomization air to about 50 *at the gun.* Then spray a test pattern.

TESTING THE SPRAY PATTERN

After setting the fan, fluid and air pressure, test the spray pattern on a piece of masking paper (Figure 3–22). Hold the gun eight to ten inches away from the masking paper. Pull the trigger all the way back and release it immediately. This burst of paint should leave a long, slender pattern on the test paper (Figure 3–23). Turn the fan control knob in until the spray pattern is six to eight inches wide.

For spot repair, the pattern should be about five to six inches from top to bottom. Adjust the fluid control to prevent flooding. For panel or overall repair, the length of the pattern should be about nine inches from top to bottom. A larger pattern can be obtained by opening the fan control knob.

Carefully inspect the texture of the spray pattern (Figure 3–24). If the paint droplets are coarse and large, turn the fluid control knob in about one-half turn or increase the air pressure five pounds. If the spray is too fine or too dry, either open the fluid control knob about one-half turn or decrease the air pressure five pounds. Repeat until the desired texture is obtained.

SHOP TALK _____

Remember the objective of spraying on a test surface is twofold. First, make sure all atomized paint particles are of uniform size. Second, make sure this size is fine enough to achieve proper flowout (Figure 3–25).

Next, test the spray pattern for uniformity of paint distribution. Loosen the air cap retaining ring and rotate the air cap (Figure 3–26) so that the horns are straight up and down. The air cap in this position will produce a horizontal spray pattern rather than a vertical one. Spray again, but this time hold the trigger

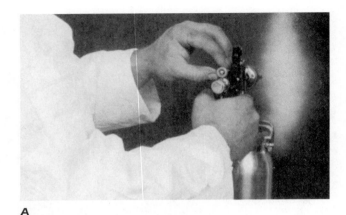

A

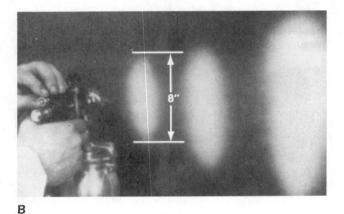

B

FIGURE 3-23 Adjust the fan control knob to get a pattern about eight inches tall.

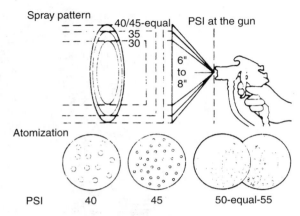

FIGURE 3-25 Notice how the particle size decreases as the air pressure increases.

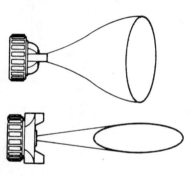

FIGURE 3-26 Rotating the air cap changes the spray pattern.

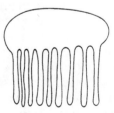

FIGURE 3-27 Balanced flood pattern.

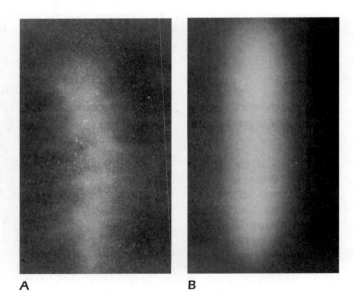

A   B

FIGURE 3-24 Inspect the spray texture: (A) dry (B) wet

down until the paint begins to run. This is called *flooding* the pattern. Inspect the lengths of the runs. If all adjustments are correct, the runs will be approximately equal in length (Figure 3-27).

The uneven runs in the split pattern shown in Figure 3-28 are a result of setting the spray pattern too wide or the air pressure too high. Turn the pattern control knob in one-half turn or raise the air pressure

FIGURE 3–28   Split flood pattern.

FIGURE 3–29   Heavy center flood pattern.

FIGURE 3–31   A hand can be used to quickly determine where to hold the spray gun.

SPRAY GUN HANDLING

Spray gun adjustments and spray gun handling work together to make a beautiful painted surface. Spray gun handling is a balance of uniform distance from the surface and speed of movement along the surface. To achieve consistent results, if you change distance you must change speed.

SHOP TALK _____

Always test the spray gun before painting. The masking paper on the vehicle can be sprayed to check the spray gun adjustments.

_____

Flat Panel

1. Hold the spray gun the proper distance from the surface—about eight to ten inches or a hand's length (Figure 3–31). If the gun is held too close to the surface, the air pressure will ripple the paint as it is applied, resulting in runs. If the gun is held too far away, the paint will dry before it reaches the surface. This will result in a dull finish and excessive overspray (Figure 3–32).
2. The air cap must be kept perpendicular to the surface. On a flat panel, this means that you must bend your wrist (Figure 3–33). It seems unnatural to bend the wrist so much, but it will insure that the paint pattern evenly strikes the surface.

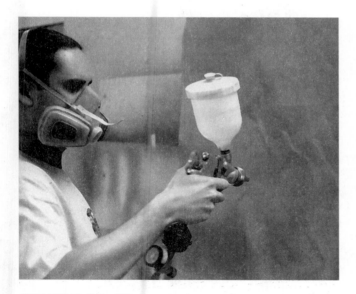

FIGURE 3–30   Wear a respirator when test spraying.

five pounds. Alternate between these two adjustments until the runs are even in length.

If paint runs are longer in the middle than on the edges (Figure 3–29), too much paint is being discharged. Turn the fluid control knob in until the runs are even in length.

_____

WARNING: Always wear a suitable air respirator (Figure 3–30) when spraying test patterns.

_____

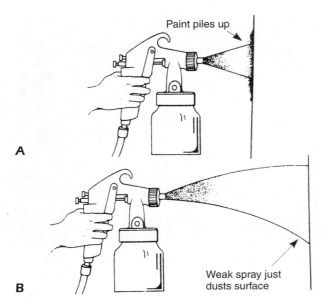

Paint piles up

A

Weak spray just
dusts surface

B

FIGURE 3–32   (A) The spray gun is too close,
resulting in runs. (B) The spray gun is too far away,
resulting in dry spray.

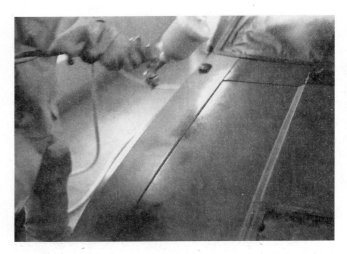

FIGURE 3–34   Start spraying at one edge of a hood
and work toward the center.

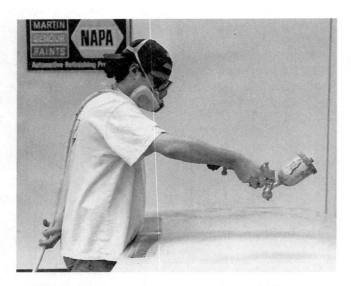

FIGURE 3–33   To keep the spray gun pattern equal
on a flat surface, the painter's wrist must bend.

3. The gun should be in motion before the trigger
   is pulled and the trigger should be released
   before the gun motion stops. This technique
   gives a "fade in" and "fade out" effect, which
   prevents overloading where one series of
   strokes is joined to the next by overlapping the
   stroke ends.

4. Move about one foot per second. Watch the
   surface as the paint is applied. A balance of
   speed and distance will provide proper paint
   application. If the paint is to dry, slow down
   slightly or move in one inch closer to the
   surface. If the paint is too wet, speed up
   slightly or move away one inch.

5. On hoods, deck lids, and roofs, start painting at
   the edge of one side. Work toward the center
   (Figure 3–34). When you reach the center,
   quickly move to the other side. Start at the wet
   edge at the center and work from the center to
   the edge.

6. Always keep the air hose over your shoulder
   (Figure 3–35). This keeps it out of the wet
   paint. Also watch your paint suit when you lean
   over to reach the center of a hood.

7. On large panels, hold your arm straight and
   walk the length of the panel. This gives a more
   uniform surface than if you were to stand in
   one spot and overlap the spray areas.

8. One problem when painting a flat surface is
   paint flow. The paint should be applied wet
   enough to allow the paint to flow out and
   smooth out before it dries.

9. Release the trigger at the end of each pass.
   Then pull back the trigger when beginning the
   pass in the opposite direction. In other words,
   "trigger" the gun and turn off the gun at the end
   of each sweep. This avoids runs, minimizes
   overspray, and saves paint. Proper triggering
   involves four steps: (1) Begin the stroke over
   the masking paper, triggering the gun halfway

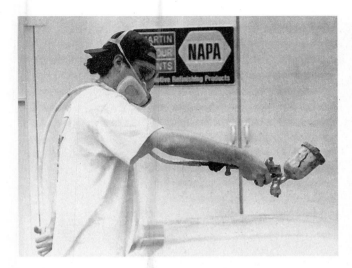

FIGURE 3-35 The air hose over the painter's shoulder keeps the hose out of the wet paint.

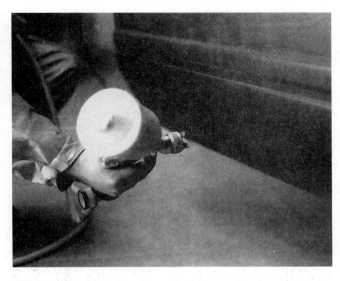

FIGURE 3-37 Begin spraying at the bottom of a panel.

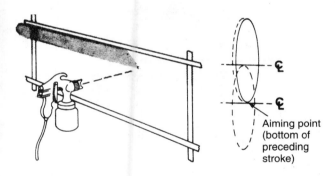

FIGURE 3-36 This is a 50 percent overlap.

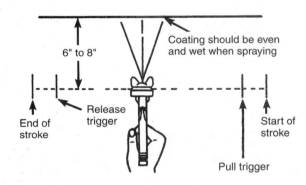

FIGURE 3-38 The paint gun should be held parallel to the surface being painted.

to release only air. (2) When the starting edge of the panel is reached, squeeze the trigger all the way to release the paint. (3) Release the trigger halfway to stop the paint flow when directly over the finishing edge. (4) Continue the stroke several more inches before reversing the direction and repeating the sequence.

10. Difficult areas such as corners and edges should be sprayed first. Aim directly at the area so that half of the spray covers each side of the edge or corner. Hold the gun an inch or two closer than usual, or screw the pattern control knob in a few turns. Either technique will reduce the pattern size. If the spray gun is just held closer, the stroke will have to be faster to compensate for a normal amount of material being applied to the smaller areas. After all of the edges and corners have been sprayed, the flat or nearly flat surfaces should be sprayed.

11. Use a 50 percent overlap. Each successive coat covers half of the previous coat (Figure 3-36). This can be accomplished by pointing the air cap at the top or bottom of the dry paint edge from the previous pass.

Vertical Panel

1. Hold the spray gun at the proper distance. The painter may start at the top or bottom of the panel. However if the painter starts at the bottom of the panel and there is an initial problem with the spray gun, paint, or application technique, the problem will be in an inconspicuous area (Figure 3-37).

2. Keep the air cap perpendicular to the surface at all times (Figures 3-38 and 3-39). You will need to squat or bend over when painting lower parts. If there is not enough room to hold the gun upright, tilt it to the side (Figure 3-40).

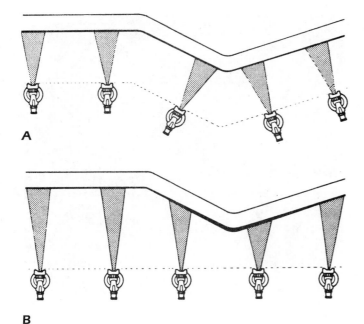

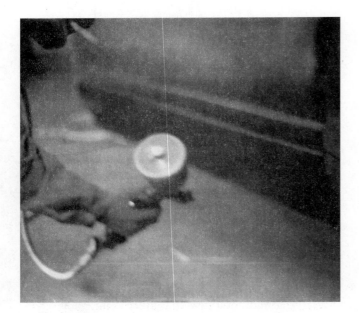

FIGURE 3–39 (A) Always keep the gun parallel even when the surface is angled. (B) Not keeping the gun parallel leads to excessive paint buildup and runs.

FIGURE 3–40 Tilt the gun to keep the pattern parallel when spraying a rocker panel.

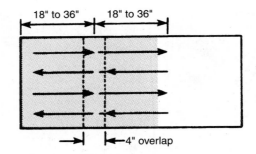

FIGURE 3–41 Keep the edges wet.

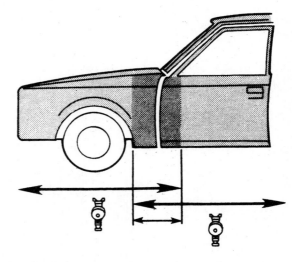

FIGURE 3–42 Switch overlap areas on successive coats.

3. If you are painting a single panel, such as a fender, position yourself just a little off-center and stretch to reach all areas with controlled, even strokes.

4. If you are painting the entire side of a vehicle, you can walk the entire side while holding your arm straight. Keep the air hose out of the way. Another method is to paint the side in overlapping sections. Usually a single panel is painted at a time. The adjacent panel is painted, and the wet edge of the first panel is blended into, by arcing the gun (Figure 3–41). Be careful of runs in this area. Successive coats should have different overlap areas (Figure 3–42).

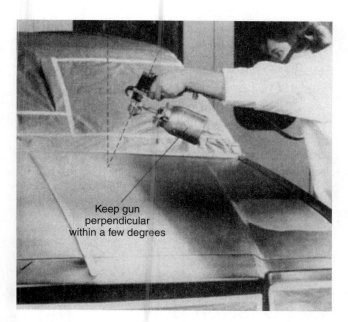

Keep gun
perpendicular
within a few degrees

FIGURE 3–43   Heeling. This painter is not keeping the fan parallel to the surface. Uneven paint build-up results from this.

5. On vertical panels, gravity pulls the paint down, increasing the flowout. Excessive coverage and resultant runs are always a possibility. Constantly monitor the speed and distance to prevent excessive coverage.

Gun Handling Problems

The inexperienced painter is prone to several spraying errors, including:

- **Heeling.** This occurs when the painter allows the gun to tilt (Figure 3–43). Because the gun is no longer perpendicular to the surface, the spray produces an uneven layer of paint, excessive overspray, dry spray, and orange peel.
- **Arcing.** This occurs when the gun is not moved parallel with the surface (Figure 3–44). At the outer edges of the arced stroke, the gun is farther away from the surface than at the middle of the stroke. The result is uneven film buildup, dry spray, excessive overspray.
- **Speed of stroke.** If the stroke is made too quickly, the paint will be dry and not cover the surface evenly (Figure 3–45). If the stroke is made too slowly, sags and runs will develop (Figure 3–46). The proper stroking speed is something that comes with experience.

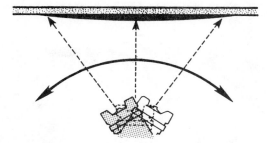

FIGURE 3–44   Arcing. Notice how uneven paint buildup results when the gun is not moved parallel to the surface.

FIGURE 3–45   Dry spray results from gun movement that is too fast.

FIGURE 3–46   Runs result from slow gun movement.

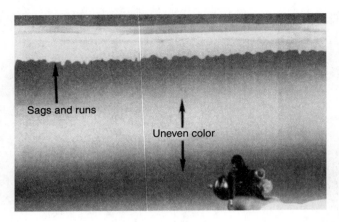

FIGURE 3–47 *Runs and improper coverage result from improper overlap.*

FIGURE 3–49 *Improper coverage at the top.*

FIGURE 3–48 *Wasteful overspray.*

- **Improper overlap.** Improper overlapping results in uneven film thickness, contrasting color hues, and sags and runs as shown in Figure 3–47.
- **Wasteful overspray.** Failure to trigger the gun before and after each stroke results in wasteful overspray and excessive buildup of paint at the beginning and end of each stroke (Figure 3–48).
- **Improper coverage.** Triggering at the wrong time is another common error. Failure to trigger exactly over the edge of the panel results in uneven coverage and film thickness (Figure 3–49).

TYPES OF SPRAY COATS

There are varying degrees of thickness for a sprayed coat. Generally, they are referred to as light, medium, or heavy. The easiest way to control this degree of thickness is by the speed with which the gun is moved. That is, the slower the speed, the heavier the coat.

There are also five other terms to describe spray coats:

- Fog coat
- Medium wet coat
- Full wet coat
- Blending coat
- Banding coat

A fog coat is used to apply a thin-bodied material such as clear coat adhesion promoter. To prevent runs in thin-bodied material, a fog coat utilizes faster-than-normal gun movement.

A medium wet coat is used quite often when refinishing vehicles. The speed of movement is normal, about 1 foot per second. The gun is kept 8–10 inches away from the surface.

A full wet coat is a heavy, glossy coat that is applied in a thickness almost heavy enough to run. More paint is applied than in a medium wet coat; either the spray gun is held closer to the surface or the spray gun is moved slightly slower.

Blending coats are applications of paint on the boundary of a spot or panel repair. The first coat covers the undercoat, the second coat extends a few inches beyond the first, and the third extends a few inches beyond the second.

Banding is a single coat applied in a small spray pattern to the frame in an area to be sprayed. This technique assures the painter of coverage at the edges without spraying beyond the spray area, and it reduces overspray. Banding is often used in spraying panel repairs with a surfacer.

Sometimes a banding coat is diluted more than the normal application that follows. This is especially true when the paint to be sprayed is of high viscosity. The additional dilution of the paint used for banding allows it to fully enter cracks and seams. A good example is the application of textured vinyl finish. Table 3–3 summarizes the variables that control quality when spray painting.

3.6 CLEANING THE SPRAY GUN

Neglect and carelessness are responsible for the majority of spray gun difficulties. Proper care of a gun requires little time and effort. Thorough cleaning of gun and accessory equipment _immediately after use,_ lubrication of bearing surfaces and packings at recommended intervals, and proper care in handling (do not drop or throw gun) are important factors in the care of a spray gun.

To clean a spray gun, a spray gun cleaner may be required by local or state law. Check on the requirement. If a spray gun cleaner is not required, the following recommendations may be used to clean spray guns.

To clean a siphon-feed spray gun, remove the cup from the gun. Empty the cup into the hazardous waste barrel. Rinse out the cup with clean thinner. Empty the cup again and refill with about 1 inch of thinner. Take the spray gun back into the booth, wear all proper safety equipment, and reconnect the air hose. Turn on the exhaust fan. Spray the thinner through the gun.

On pressure-feed air guns, release the pressure in the cup, loosen the air nozzle, and force the material into the cup by triggering the gun. Take the air nozzle off. Empty the contents into a suitable container and refill the cup with a clean compatible solvent. The air nozzle can be left off. Spray solvent from the gun and repeat this process until clean solvent is flowing from the gun. If a tank is used, clean as directed by the manufacturer and reassemble for future use.

With a gravity-feed gun, empty the cup and rinse it with a clean thinner. Then pour it out and add more

TABLE 3–3:	SUMMARY OF VARIABLE CONTROLLING QUALITY IN SPRAY FINISHING
Atomization	1. Fluid viscosity 2. Air pressure 3. Fan pattern width 4. Fluid velocity or fluid pressure 5. Fluid flow rate 6. Distance of spray gun from work
Evaporation stages	1. Between spray gun and part 2. From sprayed part
Evaporation variables between spray gun and sprayed part	1. Type of reducing thinner 2. Atomization pressure 3. Amount of thinner 4. Temperature in spray area 5. Degree of atomization

Evaporation variables affecting	1. Physical properties of solvents (i.e., fast or slow evaporation)
	2. Temperature
	  a. Fluid
	  b. Work
	  c. Air
	3. Exposed area of the surface sprayed

Evaporation variables from the sprayed part	1. Surface temperature 2. Room air temperature 3. Air pressure velocity 4. Flash time between coats 5. Flash time after final coats 6. Physical properties of the solvents (i.e., fast or slow evaporation)

Operator variables	1. Distance of spray gun from the work surface
	2. Stroking speed over the work surface
	3. Pattern overlap
	4. Spray gun attitude
	  a. Heeling
	  b. Arcing
	  c. Fanning
	5. Triggering

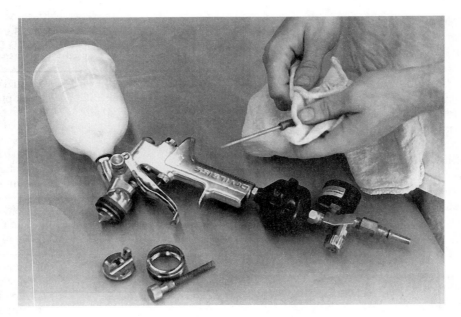

FIGURE 3–50 *Gravity-feed gun cleaning.*

clean thinner. Pull the trigger and release the thinner into the waste barrel. Remove the air cap, fluid needle, and cup lid and clean with a rag (Figure 3–50). To reduce cleaning, some types of gravity-feed spray gun cups have disposable plastic liners. The plastic liner is inserted into the cup and the paint is poured in. After spraying is completed, the liner is removed. The only part of the gun in need of cleaning is the internal paint passage, a very small area.

With any type of spray gun, remove the air cap and soak it in thinner or solvent. Clean out clogged holes with a soft material such as a round toothpick or a broom straw. Remember, *never* use wires or nails to clean the precision-drilled openings. Clean the fluid tip with a gun brush and solvent. With a clean rag soaked in thinner, wipe the outside of the gun to remove all traces of paint.

Areas in the United States with air pollution problems, such as southern California, require the use of enclosed spray gun cleaning equipment. The paint spraying equipment—guns, cups, stirrers, and strainers—is placed in the larger tub of the gun washer/recycler, the lid is closed, then the air-operated pump recirculates the thinner into the upper portion of the tub, cleaning the inside and outside of the paint gun and other pieces. In less than sixty seconds, the equipment is clean and ready for use (Figure 3–51).

FIGURE 3–51 Siphon-feed guns can be cleaned inside a paint gun washer.

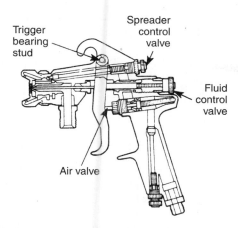

FIGURE 3-52 Parts to be lubricated on a siphon-feed gun.

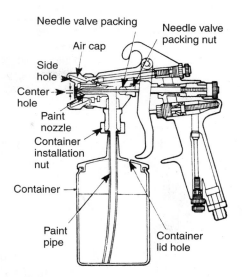

FIGURE 3-53 Potential trouble spots on a siphon-feed gun.

The gun washer/recycler saves the painter time. Compared with traditional manual gun-cleaning methods, the automatic gun washer/recycler machine saves ten minutes on each color change. The cleaning system offers increased safety for the painter, because the skin no longer is exposed to the drying effects of solvent. The system is designed so sludge from the cleaning action settles to the bottom for easy drainage and disposal with other shop wastes. Check the owner's manual for complete operational details and proper solvents to use.

SHOP TALK

If the gun is not cleaned soon after use, the fluid tip might clog completely or partially, causing the gun to split (eject pieces of dried paint) or form the wrong spray pattern. For enamel paints with additives, this is of particular importance, because the paint will harden right in the gun if not properly cleaned out.

Most spray gun manufacturers recommend lubricating, at the end of each day, the parts shown in Figure 3–52 with light machine oil. Packings and springs plus needles and nozzles must periodically be replaced due to normal wear and tear. This should be done only in accordance with the manufacturer's instructions. Extreme care should be taken not to overlubricate, since the excess oil could overflow into the paint and oil passages, mixing with the paint and resulting in a defective paint film. Oil and paint do not mix to produce a good finish.

SHOP TALK

Never soak the entire gun in cleaning solvent. Doing this will dry out the packings and remove lubrication.

For best results in refinishing, use separate spray guns for topcoats and undercoats. In fact, in any shop where there are sufficient jobs going through to justify it, there should be at least three guns, one gun to be used primarily for spraying undercoats like surfacers, another for spraying basecoats, and a third for spraying clear coats. If these guns are kept clean and in good working order, much time will be saved over trying to make one gun serve and having to adjust it each time the operation is changed.

3.7 SPRAY GUN TROUBLESHOOTING

If the spray gun is not adjusted, manipulated, and cleaned properly, it will apply a defective coating to the surface. Fortunately, defects from incorrect handling and improper cleaning can be tracked down quite readily and then corrected without much difficulty. The most common spray gun application problems, with their possible causes and suggested remedies, are given in Chapter 12.

If not properly maintained, the spray gun itself (Figure 3–53) can also create some problems. Table 3–4 contains the causes and possible solutions to some of more common spray gun difficulties.

Failure of the compressed air supply system to perform properly can cause the paint problems as shown in Table 3–5.

TABLE 3–4: TROUBLESHOOTING AN AIR SPRAY GUN

Trouble	Possible Cause	Suggested Correction
Spray pattern top-heavy or bottom-heavy	1. Horn holes partially plugged (external mix). 2. Fluid tip clogged, damaged, or not installed properly. 3. Dirt on air cap seat or fluid tip seat.	1. Remove air cap and clean. 2. Clean, replace, or reinstall fluid tip. 3. Remove and clean seat.
Spray pattern heavy to right or to left	1. Air cap dirty or orifice partially clogged. 2. Air cap damaged. 3. Paint nozzle clogged or damaged. 4. Too low a setting of the pattern control knob.	1. To determine where buildup occurs, rotate cap 180 degrees and test spray. If pattern shape stays in same position, the condition is caused by fluid buildup on fluid tip. If pattern changes with cap movement, the condition is in the air cap. Clean air cap, orifice, and fluid tip accordingly. 2. Replace air cap. 3. Clean or replace paint nozzle. 4. Adjust setting.
Spray pattern heavy at center	1. Atomizing pressure too low. 2. Fluid of too great viscosity. 3. Fluid pressure too high for air cap's normal capacity (pressure feed). 4. Caliber of paint nozzle enlarged due to wear. 5. Center hole enlarged.	1. Increase pressure. 2. Thin fluid with suitable thinner. 3. Reduce fluid pressure. 4. Replace paint nozzle. 5. Replace air cap and paint nozzle.
Spray pattern split	1. Not enough fluid. 2. Air cap or fluid tip dirty. 3. Air pressure too high. 4. Fluid viscosity too thin.	1. Reduce air pressure or increase fluid flow. 2. Remove and clean. 3. Lower air pressure. 4. Thick fluid viscosity.
Pinholes	1. Gun too close to surface. 2. Fluid pressure too high. 3. Fluid too heavy.	1. Stroke 6 to 8 inches from surface. 2. Reduce pressure. 3. Thin fluid with thinner.
Orange peel (surface looks like orange peel	1. Too high or too low an atomization pressure. 2. Gun too far from or too close to work. 3. Fluid not thinned. 4. Improperly prepared surface. 5. Gun stroke too rapid. 6. Using wrong air cap. 7. Overspray striking a previously sprayed surface. 8. Fluid not thoroughly dissolved.	1. Correct as needed. 2. Stroke 6 to 8 inches from surface. 3. Use proper thinning process. 4. Surface must be prepared. 5. Take deliberate, slow stroke. 6. Select correct air cap for the fluid and feed. 7. Select proper spraying procedure. 8. Mix fluid thoroughly.
Excessive spray fog or overspray	1. Atomizing air pressure too high or fluid pressure too low. 2. Spraying past surface of the product. 3. Wrong air cap or fluid tip. 4. Gun stroked too far from surface. 5. Fluid thinned out too much.	1. Correct as needed. 2. Release trigger when gun passes target. 3. Ascertain and use correct combination. 4. Stroke 6 to 8 inches from surface. 5. Add correct amount of thinner.

TABLE 3–4:   TROUBLESHOOTING AN AIR SPRAY GUN (Continued)

Trouble	Possible Cause	Suggested Correction
No control over size of pattern	1. Air cap sealed is damaged. 2. Foreign particles are lodged under the seal.	1. Check for damage, replace if necessary. 2. Make sure surface that this sets on is clean.
Sags or runs	1. Dirty air cap and fluid tip. 2. Gun manipulated too close to surface. 3. Not releasing trigger at end of stroke (when stroke does not go beyond object). 4. Gun manipulated at wrong angle to surface. 5. Fluid piled on too heavy. 6. Fluid thinned out too much. 7. Fluid pressure too high. 8. Operation too slow. 9. Improper atomization.	1. Clean cap and fluid tip. 2. Hold the gun 6 to 8 inches from surface. 3. Release trigger after every stroke. 4. Work gun at right angles to surface. 5. Learn to calculate depth of wet film of fluid. 6. Add correct amount of fluid by measure. 7. Reduce fluid pressure with fluid control knob. 8. Speed up movement of gun across surface. 9. Check air and fluid flow; clean cap and fluid tip.
Streaks	1. Dirty or damaged air cap and/or fluid tip. 2. Not overlapping strokes correctly or sufficiently. 3. Gun moved too fast across surface. 4. Gun held at wrong angle to surface. 5. Gun held too far from surface. 6. Air pressure too high. 7. Split spray. 8. Pattern and fluid control not adjusted properly.	1. Same as for sags. 2. Follow previous stroke accurately. 3. Take deliberate, slow strokes. 4. Same as for sags. 5. Stroke 6 to 8 inches from surface. 6. Use least air pressure necessary. 7. Reduce air adjustment or change air cap and/or fluid tip. 8. Readjust.
Gun sputters constantly Sputtering spray	1. Connections, fittings, and seals loose or missing. 2. Leaky connection on fluid tube or fluid needle packing (suction gun). 3. Lack of sufficient fluid in container. 4. Tipping container at an acute angle. 5. Obstructed fluid passageway. 6. Fluid too heavy (siphon feed). 7. Clogged air vent in canister top (siphon feed). 8. Dirty or damaged coupling nut on canister top (siphon feed). 9. Fluid pipe not tightened to pressure tank lid or pressure cup cover. 10. Strainer is clogged up. 11. Packing nut is loose. 12. Fluid tip is loose. 13. O-ring on tip is worn or dirty. 14. Fluid hose from paint tank loose. 15. Jam nut gasket installed improperly or jam nut loose.	1. Tighten and/or replace as per owner's manual. 2. Tighten connections; lubricate packing. 3. Refill container with fluid. 4. If container must be tipped, change position of fluid tube and keep container full of fluid. 5. Remove fluid tip, needle, and fluid tube and clean. 6. Thin fluid. 7. Clean. 8. Clean or replace. 9. Tighten; check for defective threads. 10. Clean strainer. 11. Make sure packing nut is tight. 12. Tighten fluid tip. Torque to manufacturer's specifications. 13. Replace O-ring if necessary. 14. Tighten. 15. Inspect and correctly install or tighten nut.

continued

TABLE 3–4:   TROUBLESHOOTING AN AIR SPRAY GUN (Continued)

Trouble	Possible Cause	Suggested Correction
Uneven spray pattern	1. Damaged or clogged air cap. 2. Damaged or clogged fluid tip.	1. Inspect air cap and clean or replace. 2. Inspect fluid tip and clean or replace.
Fluid leaks from spray gun Nozzle drip	1. Fluid needle packing not too tight. 2. Fluid needle packing dry. 3. Foreign particle blocking fluid tip. 4. Damaged fluid tip or fluid needle. 5. Wrong fluid needle size. 6. Broken fluid needle spring.	1. Loosen nut; lubricate packing. 2. Lubricate needle and packing frequently. 3. Remove tip and clean. 4. Replace both tip and needle. 5. Replace fluid needle with correct size for fluid tip being used. 6. Remove and replace.
Fluid leaks from packing nut Packing nut leak	1. Loose packing nut. 2. Packing is worn out. 3. Dry packing.	1. Tighten packing nut. 2. Replace packing. 3. Remove and soften packing with a few drops of light oil.
Fluid leaks through fluid tip when trigger is released	1. Foreign particles lodged in the fluid tip. 2. Fluid needle has paint stuck on it. 3. Fluid needle is damaged. 4. Fluid tip has been damaged. 5. Spring left off fluid needle.	1. Clean out tip and strain paint. 2. Remove all dried paint. 3. Check for damage; replace if necessary. 4. Check for nicks; replace if necessary. 5. Make sure spring is replaced on needle.
Excessive fluid	1. Not triggering the gun at each stroke. 2. Gun at wrong angle to surface. 3. Gun held too far from surface. 4. Wrong air cap or fluid tip. 5. Depositing fluid film of irregular thickness. 6. Air pressure too high. 7. Fluid pressure too high. 8. Fluid control knob not adjusted properly.	1. It should be a habit to release trigger after every stroke. 2. Hold gun at right angles to surface. 3. Stroke 6 to 8 inches from surface. 4. Use correct combination. 5. Learn to calculate depth of wet film of finish. 6. Use lease amount of air necessary. 7. Reduce pressure. 8. Readjust.
Fluid will not come from spray gun	1. Out of fluid. 2. Grit, dirt, paint skin, etc., blocking air gap, fluid tip, fluid needle, or strainer. 3. No air supply. 4. Internal mix cap using siphon feed.	1. Add more spray fluid. 2. Clean spray gun thoroughly and strain spray fluid; always strain fluid before using it. 3. Check regulator. 4. Change cap or feed.
Fluid will not come from fluid tank or canister	1. Lack of proper air pressure in fluid tank or canister. 2. Air intake opening inside fluid tank or canister clogged by dried-up finish fluid. 3. Leaking gasket on fluid tank cover or canister top. 4. Gun not converted correctly between canister and fluid tank. 5. Blocked fluid hose. 6. Connections with regulator not correct.	1. Check for air leaks or leak of air entry; adjust air pressure for sufficient flow. 2. This is a common trouble; clean opening periodically. 3. Replace with new gasket. 4. Correct per owner's manual. 5. Clear. 6. Correct as per owner's manual.

TABLE 3–4:   TROUBLESHOOTING AN AIR SPRAY GUN (Continued)

Trouble	Possible Cause	Suggested Correction
Sprayed coat short of liquid material	1. Air pressure too high. 2. Fluid not reduced or thinned correctly. (Suction feed only) 3. Gun too far from work or out of adjustment.	1. Decrease air pressure. 2. Reduce or thin according to directions; use proper thinner or reducer. 3. Adjust distance to work; clean and adjust gun fluid and spray pattern controls.
Spotty, uneven pattern, slow to build	1. Inadequate fluid flow. 2. Low atomization air pressure. (Suction feed only) 3. Gun motion too fast.	1. Back fluid control knob to first thread. 2. Increase air pressure, prebalance gun. 3. Move at moderate pace.
Unable to get round spray	1. Pattern control knob not seating properly.	1. Clean or replace.
Dripping from fluid tip	1. Dry packing. 2. Sluggish needle. 3. Tight packing nut. 4. Spray head misaligned on type MBC guns causing needle to bind.	1. Lubricate packing. 2. Lubricate. 3. Adjust. 4. Tap all around spray head with wood and rawhide mallet and retighten locking bolt.
Excessive overspray	1. Too much atomization air pressure. 2. Gun too far from surface. 3. Improper stroking, i.e., arcing, moving too fast.	1. Reduce. 2. Check distance. 3. Move at moderate pace, parallel to work surface.
Excessive fog	1. Too much or quick drying thinner. 2. Too much atomization air pressure.	1. Remix. 2. Reduce.
Will not spray on pressure feed	1. Control knob on canister cover not open. 2. Canister is not sealing. 3. Spray fluid has not been strained. 4. Spray fluid in canister top threads. 5. Gasket in canister top worn or left out. 6. No air supply. 7. Fluid too thick. 8. Clogged strainer.	1. Set this knob for pressure spraying. 2. Make sure canister is on tightly. 3. Always strain before using. 4. Clean threads and wipe with grease. 5. Inspect and replace if necessary. 6. Check regulator. 7. Thin fluid with proper thinner. 8. Clean or replace strainer.
Will not spray on suction feed	1. Spray fluid is too thick. 2. Internal mix nozzle used. 3. Spray fluid has not been strained. 4. Hole in canister cover clogged. 5. Gasket in canister top worn or left out. 6. Plug or clogged strainer. 7. Fluid control knob adjusted incorrectly. 8. No air supply.	1. Thin fluid with thinner. 2. Install external mix nozzle. 3. Always strain before use. 4. Make sure this hole is open. 5. Inspect and replace if necessary. 6. Clean or replace strainer. 7. Correct adjustment. 8. Check regulator.

continued

TABLE 3–4: TROUBLESHOOTING AN AIR SPRAY GUN (Continued)

Trouble	Possible Cause	Suggested Correction
Air continues to flow through gun when trigger has been released (on nonbleeder guns only)	1. Air valve leaks. 2. Needle is binding. 3. Piston is sticking. 4. Packing nut too tight. 5. Control valve spring left out.	1. Remove valve, inspect for damage, clean valve, and replace if necessary. 2. Clean or straighten needle. 3. Clean piston, check O-ring, and replace if necessary. 4. Adjust packing nuts. 5. Make sure to replace this spring.
Air leak at canister gasket	1. Canister not sealing on canister cover.	1. Check gasket, clean threads, and tighten canister.
Leak at setscrew in canister top	1. Screw not tight. 2. Damaged threads on setscrew.	1. Clean threads and tighten screw. 2. Inspect and replace if necessary.
Leak between top of canister cover and gun body	1. Retainer nut is not tight enough. 2. Gasket or gasket seat damaged.	1. Check nut to make sure it is tight. 2. Inspect, clean, and replace if necessary.
Pressure Fluid Tank Problems		
Leaks air at the top of the tank lid	1. Gasket not seating properly or damaged. 2. Wing screws not tight enough. 3. Fittings leak. 4. Air pressure too high.	1. Drain off all of the air from fluid tank, thus allowing the gasket to seat. Retighten wing nuts, and fill with air again. Lid will seat tightly. 2. Make sure all wing screws are tight. By following remedy #1 (above), wing screws can be pulled down even tighter. 3. Check all fittings and apply pipe dope if necessary. 4. Maximum 60 psi. Normal w.p. 25–30 psi.
No fluid comes through the spray gun	1. Not enough pressure in tank. 2. Out of fluid. 3. Fluid passages clogged.	1. Increase regulator setting until fluid flows; do not exceed 60 psi. 2. Check fluid supply. 3. Check tube, fittings, hose, and spray gun. Clean out fittings, hose, tube, and spray gun, making sure all residual fluid is removed.

TABLE 3–5: TROUBLESHOOTING A COMPRESSED AIR SUPPLY

Fault	Result	Blistering	Nondrying	Poor Adhesion	Contamination	Poor Atomization	Poor Flow	Overloading	Sags	Popping	Slow Applications	Off-shade Metallic	Uneven Application	Dry Spray	Dirt	Remedy
Oil/water not adequately condensed out.	Oil/water at spray gun	A	C	A	C											Ensure regular drainage of air receiver, separator, and transformer. Site transformers of adequate capacity in cool places. Lubricate compressors with recommended grade of mineral oil of good emulsifying properties.
Long air line; inadequate internal bore of air line; connectors, fittings, compressor, air transformers, and regulators of inadequate capacity.	Pressure drop					B	C	C	A	A	C	A				Ensure adequate air supply with 25 feet 3/8 inch (9mm) internal bore air line with appropriate fittings. NOTE: Reduction of viscosity to give improvement may produce other defects.
Inadequate compressor capacity. No pressure regulator. Regulator diaphragm broken.	Pressure fluctuation.								A	A	A		A	A	A	
Compressed air intake filter breached. Transformer filter not properly maintained. Compressor sited in dusty area.	Dirt in compressed air														A	

A Most likely failure to be associated with the fault   B Likely failure   C Failure less likely to be associated with the fault
D Will affect color of metallics   E Health hazard

SPRAY GUN REBUILDING

Spray gun parts such as gaskets, O rings, packings, and springs are subject to wear. For maximum efficiency, spray guns need to be rebuilt when these items wear out. Some symptoms of wear include paint leaks from the paint cup or when air bubbles are found in the paint cup. Rebuild kits with instructions are available from the manufacturer for each type of spray gun. Virtually any part on the spray gun can be replaced if needed.

OTHER TYPES OF SPRAY GUNS

There are three other types of spray guns that can be found in some collision repair facilities. They are the airless spray gun, the electrostatic system, and the airbrush.

AIRLESS SPRAY GUN SYSTEM

Airless spraying equipment uses hydraulic pressure to atomize paint rather than air pressure. With the airless spray method, pressure is applied directly to the paint, which is injected at high speed through small holes in the nozzle and formed into a mist. Unlike the air spray method, there is less mixing of air in the paint and, consequently, less mist dispersion. Also, since the paint is pressurized directly, less energy is used for atomization so that with the same amount of power, a degree of atomization is accomplished that is several times that for air spraying. In fact, the pressure developed in airless equipment ranges from 1,500 to 3,000 psi. Actual pressure depends on the pump ratio of the equipment.

The airless system reduces overspray and rebound to a minimum, and application of the finish is much faster than with conventional atomized air. Because of higher pressures involved, the airless system can be used with paints and other materials that have a higher viscosity. However, this system of application can only be used where a fine finish is *not* required. It is often employed to apply the finishing coating in the truck fleet commercial vehicle refinishing business. It also has found a place for auto underbody and corrosion work (see Chapter 6). The so-called air-assisted airless system that uses some air to assist in the spraying operation tends to give a better finish.

Figure 3–54 shows a typical assembly of an airless system. The gun is connected to the pump with a single hose. When the gun is spraying, the pump delivers fluid under pressure adjusted by the air pressure to the pump. When the gun is not spraying, the fluid pressure and air pressure are balanced and the pump stops. The quality and economy of the finish is depen-

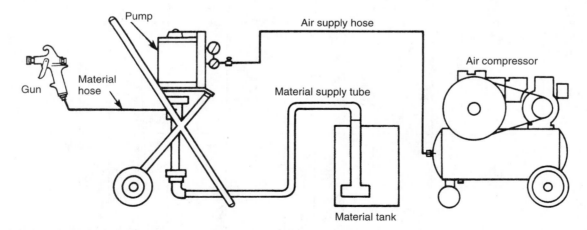

FIGURE 3–54 *Typical airless spray equipment setup.*

dent upon operator skill, fluid preparation, and nozzle size. There are six ways that a painter can control the operation of this system. They are as follows:

- **Orifice size.** This determines the amount of paint sprayed through the gun. The range for automotive coatings is 7/64 inch to 1/64 inch. More paint will be applied through the gun with a larger orifice.
- **Paint viscosity.** This is controlled by the amount of reduction. Viscosity ranges can be from 24 to 36 seconds on a #2 Zahn cup (18 to 28 seconds on a #4 Ford cup).
- **Speed of the reducer.** Generally, use the fastest reducer consistent with flow and sagging. Airless equipment sprays much wetter than conventional air-atomized equipment.
- **Speed of gun movement.** Because of the wetter spray with airless, the painter will generally have to move faster than with conventional spray equipment.
- **Gun distance.** Because of wetter spray patterns, the gun distance to the work should be around four inches.
- **Coating material.** Prepare the coating material and use the air pressure as recommended in the manufacturer's instruction manual.

The basic operating techniques of an airless spray gun are the same as those for conventional guns. That is, the gun should be held *perpendicular* and moved *parallel* to the surface in order to obtain a uniform coating of fluid. The wrist, elbow, and shoulder must all be used. Once the best working distance (ten to fifteen inches) is determined, the spray gun should be moved across the work at this optimum distance throughout the stroke.

Some object shapes do not allow this practice, but it should be used whenever possible. The proper speed allows a full wet coat application with each stroke. If the desired film thickness cannot be obtained with a single stroke or pass because of sagging, then two or more coats can be applied with a flash-off period between each coat. The spray movement should be at a comfortable rate. If the spray gun movement is excessive in order to avoid flooding the work, then the fluid nozzle orifice is too large or the fluid pressure is too high. If the stroke speed is very slow in order to apply full wet coats, then the fluid pressure should be increased slightly or a larger tip is required.

WARNING: An airless system maintains pressure after the system is shut down. High pressure can cause a serious injury. Before attempting any disassembly of the gun, system pressure must be relieved.

ELECTROSTATIC SPRAYING SYSTEM

Electrostatic spraying utilizes the principle that positive and negative electricity mutually attract each other but oppose a like charge. Therefore, when paint particles are given a negative charge by a high-voltage generator (Figure 3–55), the particles oppose each other, causing them to become atomized. On the other hand, because the adherend is grounded, it has a positive electrical charge. In this manner, when high voltage is applied between the adherend and the electrostatic painting equipment, an electrical field is formed, and the air in the field allows the electricity to pass through easily. In other words, electrical passages are formed and the atomized paint passing through these passages is sent to and adheres to the object that is being painted.

Advantages and disadvantages of electrostatic painting are as follows:

- Because the paint particles are drawn to the adherend by electrical attraction, there is less paint loss compared to normal spray painting.
- Because atomization is promoted by opposing electrical forces, a very good quality paint finish can be attained. This is particularly true for metallic painting because the metallic paint, particles are formed into rows by the opposing electrical forces, providing an appearance that

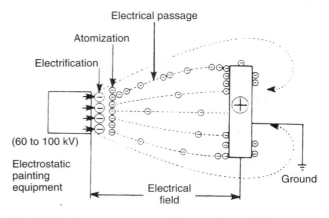

FIGURE 3–55   *Principle of electrostatic painting.*

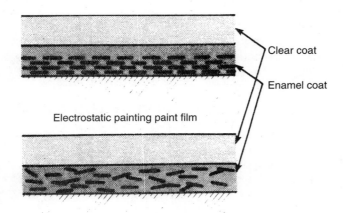

Electrostatic painting paint film

Normal air spray paint film

FIGURE 3–56 Electrostatic paint film and conventional spray film.

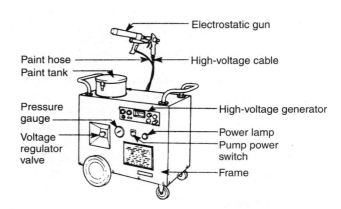

FIGURE 3–58 Airless-type electrostatic painting equipment.

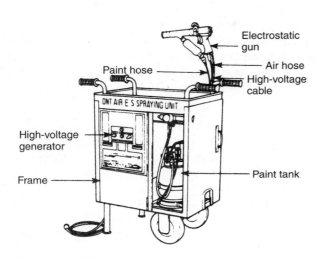

FIGURE 3–57 Air-type electrostatic painting equipment.

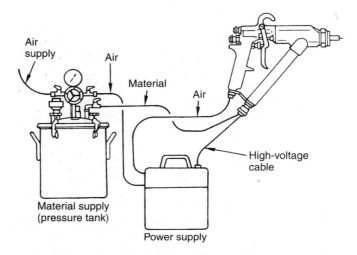

FIGURE 3–59 Atomization electrostatic equipment layout.

cannot be attained with the usual air spray gun (Figure 3–56).

- Paint adhesion efficiency is very good and, as a result, painting operations are fast. The reverse side of cylindrical objects, lattice work, and linear objects can be painted simultaneously with the front surface.
- Because the electrical potential in depressed areas is low, the adhesion is not as good, necessitating touchup.
- Unless nonconductors such as plastic, glass, and rubber are made conductive, painting is not possible.

As for portable electrostatic painting equipment, there are both the air-assisted spray type (Figure 3–57) and the airless spray type (Figure 3–58).

As with normal spray painting, a spray gun is also used for spray–type electrostatic painting, and the paint is atomized by the force of compressed air (Figure 3–59). However, atomization is further promoted by the application of a negative electrical charge. Therefore, the paint is sprayed onto the adherend by both the force of the compressed air and electrical attraction. Adhesion efficiency is not as good as with airless electrostatic spraying (Figure 3–60A), but because the spray gun is easy to use, this method is

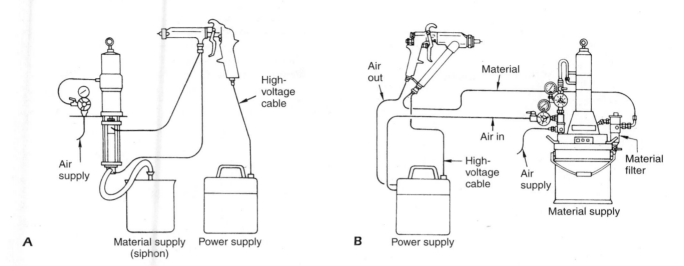

FIGURE 3-60 *(A) Airless electrostatic equipment layout; (B) air-assisted airless electrostatic equipment.*

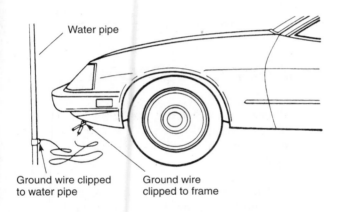

Water pipe

Ground wire clipped to water pipe   Ground wire clipped to frame

WARNING: Because of the high voltage involved in electrostatic spraying, it is very important to follow the manufacturer's instructions for the use of the equipment and all safety procedures. It is important that the vehicle be grounded. It is a good idea to always ground the car's body (or frame) to a good ground source (such as a water pipe) as soon as it enters any spray booth (Figure 3-61). Grounding the car will help prevent dust and dirt from being attracted to the new paint by static electricity.

suitable when delicate spray gun manipulation is required. The air-assisted airless electrostatic equipment overcomes this problem in some degree (Figure 3-60B).

Like the normal airless spray method, airless electrostatic spraying utilizes high pressure to atomize the paint by injecting it through small holes in the nozzle, but it also gives the paint a negative electrical charge to further promote atomization. Paint is adhered by means of both injection pressure and electrical attraction. This method provides a very good adhesion efficiency, and work is faster due to the large discharge volume. However, because compressed air is not used, injection energy is not as strong, and air spray prepainting of depressed areas like the underside of the hood and inner side of the doors is necessary.

Table 3-6 details some of the features of the six types of spray painting systems. Although at present, airless, air-type electrostatic, and airless-type electrostatic are used primarily for commercial vehicles, the environmental restrictions in some states require nearly a 65 percent transfer efficiency. The only way this seems possible is by using electrostatic, air-assisted airless, and airless techniques. Spray equipment manufacturers are hard at work trying to improve the quality of finishes obtained from these methods. In the foreseeable future, it is possible that electrostatic and airless spray equipment might be used on topcoat finishes.

AIRBRUSHES

Airbrushes range from simple types used for touchup work to complex and exacting tools used in custom finishing. The latter, of course, are generally found only in paint shops that do custom auto finishes.

TABLE 3–6:   COMPARISON OF VARIOUS SPRAYING SYSTEMS

	Normal Air Spraying	Airless Spraying	Air-Assisted Airless Spraying	Air-Assisted Electrostatic Spraying	Airless-Type Electrostatic Spraying	HVLP Spraying
Adhesion of spray efficiency	20 to 40%	50 to 60%	40 to 60%	60 to 70%	70 to 80%	65 to 90 %
Quality of finished surface	Excellent	Poor	Fair to Good	Good	Fair	Good to Excellent
Work environment (paint mist dispersion)	Poor	Good	Good	Excellent	Excellent	Excellent
Paint speed	Slow	Fast	Fast	Very fast	Very fast	Slow
Painting of depressed areas	Excellent	Poor	Fair	Good	Fair	Excellent
Gun handling (partial repainting and touch up)	Excellent	Fair	Fair	Fair	Good	Excellent

It is important to select the correct airbrush for the type of work to be performed. Consider the size and type of the work to be done, the fineness of the line desired, and the fluids to be sprayed. Airbrushes used for custom auto finishing are generally in two categories: double-action and single-action types. The double-action brushes are more versatile than are the ones found in most custom paint shops. They are available with a choice of tips to further increase their versatility. The double-action airbrush is usually recommended for projects that require very fine detailing. They produce a variable spray that works by depressing the finger-controlled front lever for air and pulling back on the same lever for the proper amount of color to be sprayed.

With single-action airbrushes, air is released by depressing the finger lever, while the amount of color desired is controlled by rotating the rear needle adjusting screw. While working, it is not possible to change the amount of color being sprayed because the operator must stop spraying to rotate the needle adjusting screw in the rear.

Airbrushes operate in a range of 5 to 50 psi pressure, with the normal operating pressure being approximately 30 psi. Compact compressors are very popular with custom auto painters.

REVIEW QUESTIONS

1. Technician A says that a gravity-feed gun cup can leak. Technician B says that a siphon-feed gun cup can leak. Who is right?
   a. technician A
   b. technician B
   c. both technician A and technician B
   d. neither technician A nor technician B
2. If there is a spray problem with the spray gun, what spray gun part is checked first?
   a. vent hole
   b. air pressure
   c. air valve
   d. fluid tip
3. Technician A says that arcing is caused by standing too far away from the vehicle when painting. Technician B says that heeling can result in runs. Who is right?
   a. technician A
   b. technician B
   c. both technician A and technician B
   d. neither technician A nor technician B

4. What type of spray gun will operate when held upside down?
   a. pressure feed
   b. gravity feed
   c. siphon feed
5. Technician A adjusts the air pressure first. Technician B adjusts the fluid first. Who is right?
   a. technician A
   b. technician B
   c. both technician A and technician B
   d. neither technician A nor technician B
6. To make the paint spray wetter, turn up the air pressure.
   a. True
   b. False
7. Technician A starts spraying at the top of the panel. Technician B starts spraying at the bottom of the panel. Who is right?

   a. technician A
   b. technician B
   c. both technician A and technician B
   d. neither technician A nor technician B
8. A _____ coat is used to spray thin bodied materials
9. Technician A cleans a gravity gun by taking out the needle and removing the air cap. Technician B cleans a siphon feed gun in a gun cleaner. Who is right?
   a. technician A
   b. technician B
   c. both technician A and technician B
   d. neither technician A nor technician B
10. HVLP spray guns have greater transfer efficiency than conventional guns.
   a. True
   b. False

Chapter

4

Compressed Air Supply Equipment

Objectives

After reading this chapter, you will be able to:
- Explain the operation of an air compressor.
- Identify the different types of filters.
- Describe the function of air and fluid control equipment.
- Properly maintain an air supply system.

Key Term List

compressor
single-stage compressor
two-stage compressor
air pressure
air transformer

The compressed air supply system is designed to provide an adequate supply of compressed air at a predetermined pressure to insure efficient operation of all air-operated equipment in the paint shop. The system can vary in size from small portable units (Figure 4–1) to large in-shop installations (Figure 4–2). The following basic requirements and considerations for these systems are the same (Figure 4–3):

- An air compressor, sometimes referred to as a *pump,* can be one compressor or a series of compressors.
- The power source is generally an electric motor. (Portable gasoline-driven compressors are available for work outside the shop.)
- A control or set of controls is used to regulate the operation of the compressor and motor.
- Air intake filters/silencers are designed to muffle intake noises as well as filter out dust and dirt.

FIGURE 4–1 Portable 3 hp 30-gallon-tank air compressor.

- The air tank or receiver must be properly sized. It cannot be too small or it will cause the compressor to cycle too often, thus causing excessive load on the motor. It should not be too large because of space problems and unnecessary capacity.

• The distribution system is the key link in the compressed air system. This is the hose or piping, or arrangement of hose and piping, from the air receiver to distribution points requiring compressed air. This distribution system consists of the proper sizes of hose or pipe, fittings, valves, air filters, oil and water extractors, regulators, gauges, lubricators, and other air and fluid control equipment that will provide for the effective and efficient operation of specific air devices, tools, and spraying equipment.

FIGURE 4–2 A paint shop air compressor setup.

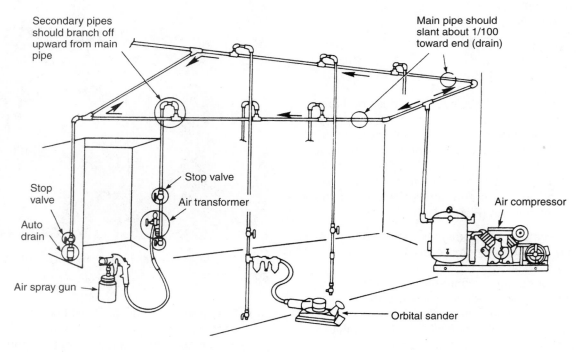

FIGURE 4–3 Typical piping arrangement found in a body/paint shop.

4.1 — AIR COMPRESSOR

The **compressor**—the heart of any air system—is designed to raise the pressure of air from normal atmospheric to some higher pressure, as measured in pounds per square inch (psi). While normal atmospheric pressure is about 14.7 psi, a compressor will typically deliver air at pressures up to 200 psi.

There are three basic types of air compressors:

- Diaphragm-type
- Piston-type
- Air screw (or rotary)–type

DIAPHRAGM-TYPE COMPRESSOR

In this type of compressor (Figure 4–4), a durable diaphragm is stretched across the bore of a very shallow compression chamber. A connecting plate, operated by an eccentric mounted on the motor shaft, alternately pulls the diaphragm down and thrusts it upward. As the diaphragm is pulled down (Figure 4–5A), air is drawn into the small space above the diaphragm. When the diaphragm is thrust upward (Figure 4–5B), the air trapped in the compression chamber is squeezed and forced out into the delivery chamber and supply lines. Although only a very small amount of air—in the 30 to 35 psi range—is compressed during each cycle, the action is very rapid—in excess of 1500 strokes per minute.

Even though most body and refinishing operations consume large quantities of compressed air at relatively higher pressures, a diaphragm compressor may be found in the average shop. The compressed air for a supplied air respirator or air brush may be provided by a diaphragm-type compressor.

PISTON-TYPE COMPRESSOR

The piston-type air compressor pump develops compressed air pressure through the action of a reciprocating piston. As shown in Figure 4–6, a piston, which is actuated by a crankshaft, moves up and down inside a cylinder, very much like a piston in an automobile engine. On the downstroke (Figure 4–6A), air is drawn into the compression chamber through a one-way valve. On the upstroke (Figure 4–6B), as the air is compressed by the rising piston, a second one-way valve opens, and the air is forced into a pressure tank or receiver. As more and more air is forced into the tank, the pressure inside the tank rises.

Piston compressor pumps are available in single- or multiple-cylinder and single- or two-stage models, depending on the volume and pressure required.

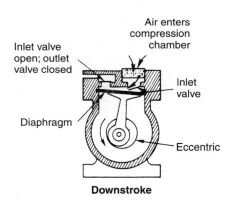

Downstroke

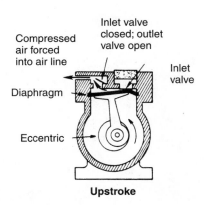

Upstroke

FIGURE 4–5 The operation of a diaphragm compressor.

FIGURE 4–4 Typical diaphragm-type compressor.

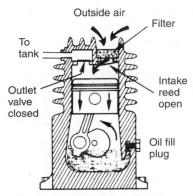

Outside air

Filter

To tank

Outlet valve closed

Intake reed open

Oil fill plug

Downstroke: Air being drawn into compression chamber

When the air is drawn from the atmosphere and compressed to its final pressure in a single stroke, the compressor is referred to as a **single-stage compressor.** Single-stage units normally are used in pressure ranges up to 125 psi for intermittent service. Most single-stage compressors are rated at 50 percent duty cycle (one-half the time on, one-half the time off). They are available in single- or multi-cylinder compressors (Figure 4–7). The principal parts of a typical piston-type compressor are shown in Figure 4–8.

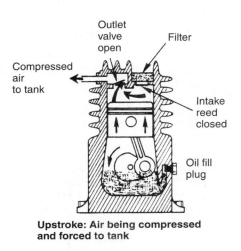

Outlet valve open

Filter

Compressed air to tank

Intake reed closed

Oil fill plug

Upstroke: Air being compressed and forced to tank

FIGURE 4–6 The operation of a piston-type compressor.

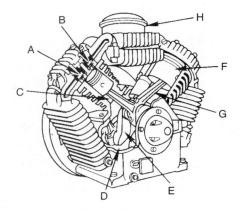

FIGURE 4–8 The principal parts of a four-cylinder, two-stage piston-type compressor: (A) intake and (B) exhaust valves, (C) second-stage piston, (D) crankcase, (E) crankshaft, (F) first-stage piston, (G) connecting rod assembly, and (H) air intake filter.

FIGURE 4–7 Single-stage compressor pumps (left to right); single cylinder, angled V-cylinders, and two cylinders.

Air drawn from the atmosphere is compressed first to an intermediate pressure and then further compressed to a higher pressure in a **two-stage compressor** (Figure 4–9). Such a compressor has cylinders of unequal bore. The first stage of compression takes place in the large-bore cylinder. In the second stage, the air is compressed for a second time to a higher pressure in the smaller bore cylinder (Figure 4–10) after passing through an intercooler. Two-stage compressors are usually more efficient, run cooler, and deliver more air for the power consumed, particularly in the 100 to 200 psi pressure range; this range of pressure is enough for most body or finishing applications.

The advantage of the piston compressor is that it is generally more durable and has a greater capacity than diaphragm types. However, since the piston rides in a cylinder, lubrication is necessary.

In recent years, an oilless or oil-free piston compression system has been introduced that employs self-lubricating materials that do not require an oil lubricant. Until recently, most oilless compressors, like the diaphragm type, were considered compacts and limited in both output and pressure. However, there are oilless compressors now on the market of up to approximately five horsepower that will nearly equal, in output and pressure, oil-lubricated compressors of the same horsepower. All oilless compressors produce a clean air output. However, they should never be used as a part of a fresh-air supplied respirator system, unless they are equipped with the necessary safety filters and controls.

ROTARY-SCREW AIR COMPRESSOR

Rotary-screw–type air compressors have been a standard in industry, but because of an oil output problem, they were never accepted by the automotive refinishing profession. Recent innovations have greatly or completely eliminated the oil problem but, except in certain areas of the country, the rotary air compressor is seldom used in auto paint shops. The rotary-screw air compressor is a highly efficient and dependable machine.

FIGURE 4–9 Typical two-stage compressor.

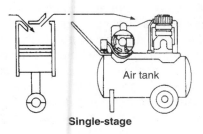

Single-stage

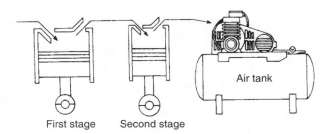

First stage   Second stage

Two-stage

FIGURE 4–10 Comparison of single- and two-stage compressors.

HOW COMPRESSORS ARE RATED

The following terms are used to measure the performance of a compressor:

- **Horsepower (hp).** Horsepower is the measure of the work capacity of the motor or engine that drives the compressor. Compressors found in body and paint shops usually range from 3 to 25 hp. As a general rule, the greater the horsepower, the more powerful the compressor.

- **Cubic feet per minute (cfm).** Cubic feet per minute is the volume of the air being delivered by the compressor to the spray gun or air tool and is used as a measure of the compressor's capabilities. Compressors with higher cfm ratings provide more air through the hose to the tool, thus making higher cfm outfits more practical for larger jobs. Actually, compressors have two cfm ratings:

  —**Displacement cfm.** This is the theoretical amount of air in cubic feet that the compressor can pump in one minute. It is a relatively simple matter to calculate the air displacement of a compressor if the piston diameter, length of stroke, and rpm are known. For example, the area of the piston multiplied by the length of the stroke and the shaft revolutions per minute equals the displacement volume. The formula for computing it is as follows:

$$\frac{\text{Area of piston} \times \text{stroke} \times \text{rpm} \times \text{number of pistons}}{1.728}$$

  —**Free air cfm.** This is the actual amount of free air in cubic feet that the compressor can pump in one minute at working pressure. The free air delivery at working pressure, not the displacement or the horsepower, is the true rating of a compressor and should be the only cfm rating considered when selecting an air compressor. Keep in mind that compressor units are frequently rated in standard cubic feet per minute (scfm), which is really cfm corrected for a given barometric pressure and temperature.

  It should be remembered that free air delivery is always less than the displacement rating, since no compressor is 100 percent efficient. The volumetric efficiency of a compressor is the ratio of free air delivery to the displacement rating, expressed as a percentage. For example, if a compressor unit for 100 pounds service has a displacement of 8 cfm and its volumetric efficiency is 75 percent, at this pressure the free air delivery will be: 8 cfm x 75 percent, or 6 cfm.

SHOP TALK

The displacement of a two-stage compressor is always given for that of the first stage cylinder or cylinders only. This is because the second stage merely rehandles the same air the first stage draws in and cannot increase the amount of air discharged.

- **Pressure (psi).** Pounds per square inch (psi) is the measure of **air pressure** or force delivered by the compressor to the air tool. This is usually expressed in two figures:
  —Normal or continuous working pressure
  —Maximum pressure
  Average psi and free air delivery requirements for various air tools and accessories are given in Chapter 2.

TANK SIZE

As previously mentioned, with most piston types of compressor pumps, air is forced into a tank or receiver. Working pressure is not available until the tank pressure is above the required psi of the air tool. The compressor puts more into the tank than is required for application. Thus, the larger the tank, the longer a job can be done at the required pressure before a pause is necessary to rebuild pressure in the tank. Since the air tank acts as a reservoir, the unit may for a short time exceed the normal capacity of the compressor. This reservoir action of the tank also reduces the running time of the compressor, thereby decreasing compressor wear and maintenance.

Air tanks or receivers usually have a cylindrical shape, and the compressor motor and pump are usually mounted on top of them. Tanks can be purchased with either horizontal (Figure 4–11A) or vertical stationary mountings (Figure 4–11B) or can be mounted horizontally on wheels for portability.

COMPRESSOR OUTFITS

As already illustrated, there are two types of compressor outfits used in paint shops: portable and stationary. A portable outfit is designed for easy movement. It is equipped with handles, wheels, or casters and usually a small air receiver or pulsation chamber.

A stationary outfit is one that is permanently installed. It is usually equipped with a larger air receiver than the portable type and might have a pressure switch as found on service station compressors or an automatic unloader as found on larger industrial units. Larger stationary models are generally equipped with a centrifugal pressure release.

A

B

FIGURE 4–11 (A) Horizontal-mounted compressor. (B) Vertical-mounted compressor.

Air filter

Belt

Motor

Pressure switch

Tank

Drain

Two-stage compressor pump

FIGURE 4–12 Parts of a stationary 5 hp 80-gallon-tank air compressor.

As shown in Figure 4–12, the typical parts of a stationary compressor are:

- Air compressor pump
- Electric or gasoline motor (powers the compressor)
- Air receiver or storage tank (holds the compressed air)
- Check valve (prevents leakage of the stored air)
- Pressure switch (automatically controls the air pressure)

also on large models:

- Centrifugal pressure release (relieves the motor of starting against a load)

- Safety valve (protects air lines and equipment against excessive pressure)
- Drain valve at the bottom of the air receiver (drains condensation)

Because of the importance of the system's safety controls, it is wise to know how they operate to protect against excessive pressure and electrical problems. The most common of these include the following:

- **Automatic unloader** is a device designed to maintain a supply of air within given pressure limits on gasoline and electrically driven compressors when it is not practical to start and stop the motor during operations. When the demand for air is relatively constant at a volume approaching the main capacity of the compressor, an unloader is recommended.

When maximum pressure in the air receiver is reached, the unloader pilot valve (Figure 4–13A) opens to let air travel through a small tube to the unloader mechanism (Figure 4–13B) and holds open the intake valve on the compressor, allowing it to run idle. When pressure drops to a minimum setting, the spring-loaded pilot automatically closes, air to the unloader is shut off, causing the intake valve to close, and the compressor resumes normal operation. Maximum and minimum pressures can be varied by resetting the pressure adjusting screw on the pilot.

- **Pressure switch** is a pneumatically controlled electric switch for starting and stopping electric motors at preset minimum and maximum

pressures. That is, this switch maintains a "cut-in" low pressure point; for example, 80 psi. When the pressure in the air receiver drops to this low point, the motor will start and the compressor will then pump up to its cut-off high pressure point, which might be 100 psi, thus breaking contact and stopping the motor; when the pressure drops to its low point, the cycle is repeated. The time to pump as shown in Table 4–1 varies by compressor size and type and cut-in/cut-out pressure.

- **Motor starter** is an electrical switch designed to provide overload protection or other necessary electrical control for starting motors of various types. The design of the switch varies with different motor sizes and current characteristics.

- **Overload protection** is usually provided on small units by fuses and on larger ones by thermal overload relays on the starting device. Relays are recommended with time delay features so that circuits will not be opened by short-duration overloads that are not harmful enough to injure the motor. Overload protection should be employed on all compressor installations, except the smaller types that are designed to operate from the standard wall socket.

- **Centrifugal pressure release** is a device that allows the motor to start up and gain momentum before engaging the load of pumping air against pressure. When the compressor slows down to stop, rotating the crankshaft more slowly, steel balls (Figure 4–14A) move toward the center, where they wedge against a cam surface, forcing the cam outward. This opens a valve

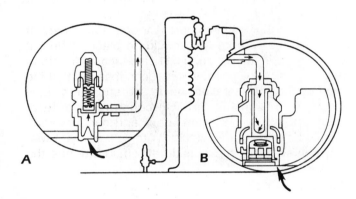

FIGURE 4–13 Automatic "pop-off" unloader. (Courtesy of ITW DeVilbiss Automotive Refinishing Products)

Outfit	HP	Cut-In Pounds Pressure	Cut-Out Pounds Pressure	Time to Pump from Cut-In to Cut-Out (in seconds)	Tank Size
Single-Stage	1	80	100	83	30 gal.
Single-Stage	2	80	100	69	60 gal.
Single-Stage	3	80	100	51	60 gal.
Two-Stage	1	140	175	284	60 gal.
Two-Stage	3	140	175	115	80 gal.
Two-Stage	5	140	175	75	80 gal.
Two-Stage	10	140	175	56	120 gal.
Two-Stage	15	140	175	42	120 gal.
Two-Stage	20	140	175	36	200 gal.
Two-Stage	25	140	175	30	200 gal.

TABLE 4–1:   TYPICAL CUT-IN/CUT-OUT COMPRESSION TIMES

(Figure 4–14B), bleeding air from the line connecting to the check valve. With air pressure bled from the pump and aftercooler, the compressor can start up free of back pressure until it gets up speed. When normal speed is reached, the balls move out by centrifugal force, releasing the cam, closing the valve, and allowing air to again be pumped into the air receiver.

- **Fused disconnect switch** is a knife-type off/on switch containing the proper size fuse. This should be used at or near the compressor unit with the line going from the fused disconnect to the starter. Fuses should be large enough to handle 2-1/2 times the current rating stamped on the motor. A qualified electrician should always make the electrical hookup of an air compressor.

4.2 — AIR AND FLUID — CONTROL — EQUIPMENT

The control of the amount, pressure, and cleanness of the air going to the pneumatic tools, especially spray guns, is of critical importance in the system. The intake air filter located on the compressor is very important, because all the air going into the compressor must pass through this filter. The filter element must be made of fine mesh or felt material to insure that small particles of grit and abrasive dust do not pass into the cylinders; thus the filter prevents excessive wear on cylinder walls, piston rings, and valves.

Once the air leaves the compressor, any equipment installed between the air pump and the point of use modifies the nature of the airstream. This modifi-

cation could be a change in pressure, in volume, in cleanliness, or some combination of them. It must be remembered that raw air piped directly from a compressor is of little use to the refinishing shop. The air contains small but harmful quantities of water, oil, dirt, and other contaminants that will lessen the quality of the sprayed finish. And the air will likely vary in pressure during the job. Furthermore, there will probably be a need for multiple air outlets for compressed air to run various pieces of equipment. Any type of item installed in the air line that performs one or more of these functions is a piece of air control equipment.

DISTRIBUTION SYSTEM

The interconnecting piping, that is, the piping from the compressor to the tool input, can be copper tubing, or galvanized or black iron pipe. Table 4–2 shows the correct pipe size in relation to compressor size and air volume.

The location of the compressor unit is important. If possible, the compressor should be placed where it can receive an ample supply of clean, cool, dry air. In most areas of the country, it is recommended that the compressor be installed inside the shop and draw clean, filtered inside air. The compressor area or room should be well ventilated, and condensation should be kept to a minimum. With cold ambient winter temperatures, the outside air could damage the compressor.

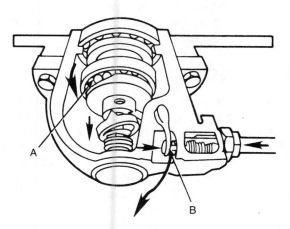

FIGURE 4–14 *Centrifugal pressure release. (Courtesy of ITW DeVilbiss Automotive Refinishing Products)*

TABLE 4–2: MINIMUM PIPE SIZE RECOMMENDATIONS*

Compressing Outfit		Main Air Line	
Size	Capacity	Length	Size
1-1/2 and 2 hp	6 to 9 cfm	Over 50 feet	3/4″
3 and 5 hp	12 to 20 cfm	Up to 200 feet	3/4″
		Over 200 feet	1″
5 to 10 hp	20 to 40 cfm	Up to 100 feet	3/4″
		Over 100 to 200 feet	1″
		Over 200 feet	1-1/4″
10 to 15 hp	40 to 60 cfm	Up to 100 feet	1″
		Over 100 to 200 feet	1-1/4″
		Over 200 feet	1-1/2″

*Piping should be as direct as possible. If a large number of fittings are used, large size pipe should be installed to help overcome excessive pressure drop.

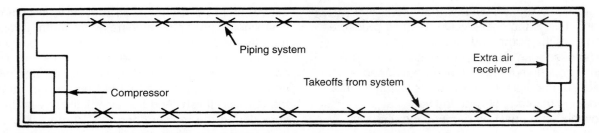

FIGURE 4–15   *Some body/paint shop layouts require an extra air receiver.*

In shops that use an outside intake, the distance between it and the compressor should be as short as possible for best efficiency, and the outside intake should be protected from the elements with a hood or suitable weatherproof shield. The compressor air intake should not be located near steam outlets or other moisture-producing areas.

The pump unit itself should be at least one foot from any wall or obstruction so that air can circulate around the compressor to aid in proper cooling. The compressor must be level. Mounting pads or vibration dampeners are generally used under the feet of the compressor. These absorb the vibration, eliminating excessive wear in the area where the feet are welded to the tank. Normally the air compressor is mounted with the flywheel facing the wall for additional safety.

The compressor should be located as near as possible to operations requiring compressed air. This cuts down lengthy air lines that cause needless pressure drop. It is good practice if the shop is long and narrow to install an extra air receiver at the far end to act as a cushion and to help reduce pressure drop when peak loads are placed on the compressed air supply.

Pressure drops, to a great extent, can be avoided by encircling the shop or looping the distribution system. This is accomplished by running the piping in a full circle or loop from the air receiver around the shop and back to the air receiver. A double loop or circle is accomplished by installing a tee in the line and then running a loop or circle in both directions back to the air tank. For this type of installation, it is recommended that an extra air tank (Figure 4–15) be installed at the far end to balance out peak loads. All piping should be installed so that it slopes toward the compressor air receiver or a drain leg installed at the end of each branch, to provide for drainage of moisture from the main air line. This line should not run adjacent to steam or hot-water piping.

In the air distribution or supply system, there should be a shut-off valve on the main line, close to the storage receiver tank. This valve is used to shut off the air at the air receiver. Keeping the air shut off at the storage tank overnight insures a full tank of air when the shop is opened each day.

FIGURE 4–16   *Air transformer.*

Other air control devices come in a very wide variety of types, but they basically all perform one or more of the following functions: air filtering and cleaning; air pressure regulation and indication of pressure; and air distribution through multiple outlets. Some typical devices to perform these functions are called air transformers, air condensers, air regulators, and in some circumstances, air lubricators.

AIR TRANSFORMER

An **air transformer,** sometimes called a *moisture separator/regulator,* is a multipurpose device that removes oil, dirt, and moisture from the compressed air, filters and regulates the air, indicates by gauge the regulated air pressure; and provides multiple air outlets for spray guns, dusters, air-operated tools, and so on. Figure 4–16 illustrates a typical air transformer. Some air transformers are equipped with a second gauge that indicates main line pressure.

Air transformers are used in all spray finishing operations that require a supply of clean, dry, regulated air. They remove entrained dirt, oil, and moisture by a series of baffles, centrifugal force, expansion cham-

FIGURE 4–17 Typical air condenser.

bers, impingement plates, and filters, allowing only clean, dry air to emerge from the outlets. The air-regulating valve provides positive control, insuring uniformly constant air pressure. Gauges indicate regulated air pressure, and in some cases, main line pressure as well. Outlets with valves allow compressed air to be distributed where it is needed. The drain valve provides for the elimination of sludge consisting of oil, dirt, and moisture. The air transformer should be installed at least twenty-five feet from the compressing unit.

AIR CONDENSER OR FILTER

An air condenser is basically a filter that is installed in the air line between the compressor and the point of use. It separates solid particles such as oil, water, and dirt out of the compressed air. No pressure regulation capability is supplied by this device. A typical air condenser is illustrated in Figure 4–17.

AIR PRESSURE REGULATOR

An air pressure regulator is a device for reducing the main-line air pressure as it comes from the compressor. It automatically maintains the required air pressure with minimum fluctuations. Regulators (Figure 4–18) are used in lines already equipped with an air condenser or other type of air filtration device. Air regulators are available in a wide range of cfm and psi capacities, with and without pressure gauges, and in different degrees of sensitivity and accuracy. They have main-line air inlets and regulated air outlets (Figure 4–19).

FIGURE 4–18 Simple air regulator.

FIGURE 4–19 Typical filter/air regulator.

FIGURE 4–20   *Air line lubricator for a paint gun cleaner.*

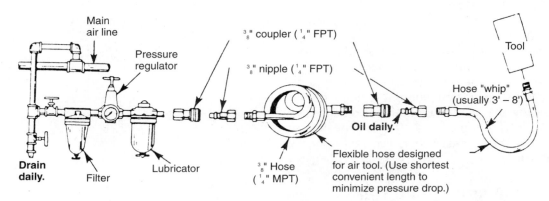

FIGURE 4–21   *Lubricator setup for air tools, not to be used to spray paint.*

LUBRICATOR

Certain types of air-operated tools and equipment described in Chapter 2 require a very small amount of oil mixed in the air supply that powers them. An automatic air line lubricator (Figure 4–20) should be installed on a leg or branch line furnishing air to pneumatic tools. (Never install a lubricator on a leg or branch air line used for paint spraying, since the oil supplied by it could damage the finish.) Lubricators are often combined with air filters and regulators in a single unit. Figure 4–21 shows a lubricator/filter/regulator unit with a built-in sight glass for determining reserve oil level.

THERMAL CONDITIONING AND PURIFICATION EQUIPMENT

While the air control devices already described in this chapter will remove contaminants from the compressed air most satisfactorily, there are special problems in some shops in the country with heat, dampness, and dirt that require special thermal conditioning and purification equipment. This equipment is usually installed between the compressor and the air storage receiver tank (Figure 4–22). It includes the following:

- **Aftercooler.** The primary purpose of this device is to reduce the temperature of compressed air.

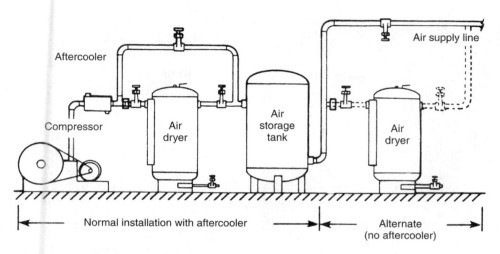

Air supply line

Aftercooler

Compressor

Air dryer

Air storage tank

Air dryer

Normal installation with aftercooler

Alternate (no aftercooler)

FIGURE 4–22 _Aftercooler setup on the left, no aftercooler setup on the right._

Heat, as well as some impurities, can be removed by installing an aftercooler in the system. Aftercoolers are very efficient in lowering air temperature and removing most of the oil and water; the residue of oil and water will be removed before it enters the air receiver. A "fish eye" condition is common in shops that do not remove the oil from their lines. There are several different designs or types of aftercoolers available. The most common is the water-cooled "air tube" design, in which air passes through small tubes and recirculating water is directed back and forth across the tubes by means of baffles and moves in a direction counter to the flow of air. This cross-flow principle is accepted as the most efficient means of heat transfer.

- **Automatic dump trap.** This trap, installed at the lowest point below the air receiver, will collect condensed moisture. It is designed so that the trap opens automatically to discharge a predetermined volume. Due to the air pressure behind the water, the trap opens and closes with a snap action that insures proper seating of the closing valve. A small line strainer should be installed ahead of any automatic device to keep foreign particles from clogging the working parts. If this unit is properly installed, with a line strainer, it will give long and satisfactory service with minimum maintenance.
- **Air dryers.** Good aftercoolers will remove the greatest percentage of water vapor, but the residue can still cause problems. To prevent such problems, an air dryer is used. There are

many designs of air dryers available; among these the most common are chemical, desiccant (a drying agent), and refrigeration types. All dryers are designed to remove moisture from the compressed air supply so that no condensation will take place in the distribution system under normal working conditions.

HOSE AND CONNECTIONS

The various types of hose used to carry compressed air and fluid to the spray gun are important parts of the system. Improperly selected or maintained hose can create a number of problems.

HOSE TYPES

There are two types of hoses in a compressed-air system: air hose and a fluid or material hose (Figure 4–23). The air hose in most compressed systems is usually covered in red rubber, although in smaller, low-pressure systems it might be covered with a black-and-orange braided fabric. The fluid or material hose is normally black or brown rubber.

SHOP TALK

The air hose is not to be used for solvent-based paints.

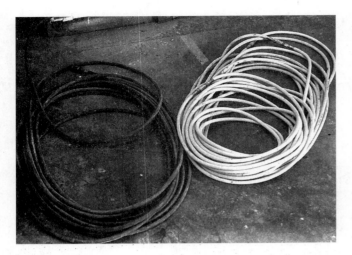

FIGURE 4–23   Fluid hose (left) and air hose (right).

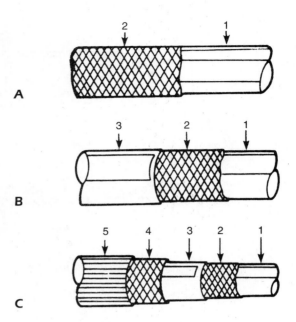

FIGURE 4–24   The construction of hoses used in compressed air systems: (A) braid-covered hose; (B) single-braid hose; and (C) double-braid hose.

The air hose is usually a simple braid-covered hose that consists of rubber tubing (1) reinforced and covered by a woven braid (2), as shown in Figure 4–24A. The single braid, rubber-covered hose (Figure 4–24B) consists of an inner tube (1), a braid (2), and an outside cover (3), all vulcanized into a single unit. The double-braid hose, illustrated in Figure 4–24C, consists of an inner tube (1), a braid (2), a separator or friction layer (3), a second layer of braid (4), and an outer rubber cover (5), all vulcanized into one. Double-braid hose can withstand a higher working pressure than single-braid hose. The inside diameter of some synthetic hoses is checked by running a small ball of the same size through it, and these are described as "ball-tested" hoses. Because of possible inner-diameter inconsistancy, untested hose may have more air pressure drop when compared to ball-tested hose.

SHOP TALK

Since the solvents in some coatings used in refinishing would readily attack and destroy ordinary rubber compounds, fluid hose is lined with special solvent-resistant material that is impervious to all common solvents.

HOSE SIZE

With both the air and fluid (material) hoses, it is important to use the proper size and type to deliver the air from its source to the air tools and guns. When compressed, air must travel a long distance; its pressure begins to drop. To minimize air pressure drop, air hoses should not be longer than twenty-five feet. Also make certain to employ only the hose constructed for compressed-air use and with a rating of at least four times that of the maximum psi being used.

TABLE 4–3:   AIR PRESSURE DROP

Size of Air Hose (ID)*	Air Pressure Drop (in PSIG)			
	5-Foot Length	15-Foot Length	25-Foot Length	50-Foot Length
1/4 Inch				
@40 PSIG†	0.4	7.5	10.5	16.0
@60 PSIG	4.5	9.5	13.0	20.5
@80 PSIG	5.5	11.5	16.0	25.0
5/16 Inch				
@40 PSIG	0.5	1.5	2.5	4.0
@60 PSIG	1.0	3.0	4.0	6.0
@80 PSIG	1.5	3.0	4.0	8.0
3/8 Inch				
@40 PSIG	1.0	1.0	2.0	3.5
@60PSIG	1.5	2.0	3.0	5.0
@80 PSIG	2.5	3.0	4.0	6.0

*ID: Inner diameter
†PSIG: Pounds per square inch gauge

Table 4–3 indicates just how much pressure drop can be expected at different pressures with hoses of varying length and internal diameters. At low pressure and with short lengths of hose, this drop is not particularly significant, but as the pressure is increased and the hose lengthened, the pressure drop rapidly be-

FIGURE 4–25 Quick-connect couplers (from left): female connector, union, male connector, and quick connect.

comes very large and must be compensated for. Too often a tool is blamed for malfunctioning when the real cause is an inadequate supply of compressed air resulting from using too small an inner diameter (ID) hose. Spray guns are entirely dependent on air pressure and volume to atomize paint. A 1/4-inch diameter air hose can not supply enough air volume to properly atomize paint. A 3/8-inch diameter air hose with 3/8-inch couplings, not longer than 25 feet, is recommended for spray gun use.

MAINTENANCE OF HOSES

A hose will last a long time if it is properly handled and maintained. Caution should be taken when it is dragged across the floor. It should never be pulled around sharp objects, run over by vehicles, kinked, or otherwise abused. Hose that ruptures in the middle of a job can ruin or delay the work.

The outside of both the air and fluid hoses should be wiped down with solvent at the end of every job, then stored by hanging up in coils.

CONNECTORS

Connections are necessary at the regulators, at the ends of hoses, and at the air tools. Of the many different types used, the most common are the threaded and quick-connect types. The former is a screw-type fitting and is usually tightened with a wrench, while the quick-connect is readily attached and detached by hand (Figure 4–25).

Both types of connections use the compression ring system to mount the fittings to the air or fluid hose. The system employs a ring that is slipped over the end of the hose. The stem on the fitting is inserted into the hose, and then a sleeve is slipped on and tightened. It forms a perfect seal against the ridges of the stem. The compression ring fitting has a number of advantages. It is economical, in that all of the parts are reusable, it forms a perfect seal free of leaks, and

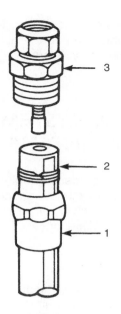

FIGURE 4–26 Compression-ring connection installation.

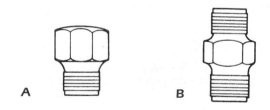

FIGURE 4–27 Typical compressed air hose line adapter (A) and coupling (B).

there is no pinching or distortion of the hose cover. The fittings are easily removed and reattached without special tools.

To install a compression ring connection, slip the sleeve (1) and the compression ring (2) over the end of the hose as shown in Figure 4–26. Hold the body (3) of the connection in a vise, and push the hose into the body as far as it will go. Slide the compression ring up to the body and bring the sleeve over the ring and thread it on by hand. Tighten it with a wrench.

ADAPTERS AND COUPLINGS

An adapter is a type of connection, shown in Figure 4–27A, that is male on one end and female on the other. It is used to convert the connections on the hose and other equipment from one thread size to another. Adapters are available in a very wide variety of sizes and threads.

A coupling is a type of connection, illustrated in Figure 4–27B, that is male on both ends. It is used to couple two pieces of hose or pipe together or to convert a female connection of one size thread to a male connection of another size thread.

4.4 AIR SYSTEM MAINTENANCE

The manufacturer's specific maintenance schedule given in the owner's manual should be followed exactly. In general, however, all air systems require the following periodic maintenance:

DAILY

- Drain the air receiver and drain the moisture separator/regulator or air transformer. If the weather is humid, drain them several times a day.
- Check the level of the oil in the crankcase. While it should be kept full, do not overfill. Overfilling causes excessive oil usage.

WEEKLY

- Pull the ring on the safety valve and unseat it. If the valve is working properly, it will release air as follows:
  —Valves located on the air receiver or check valve release air when the tank contains compressed air.
  —Valves located on the compressor's inner cooler release air only during compressor operation.
  Reseat the safety valve by pushing down on the stem with a finger. If the valve malfunctions, repair or replace immediately.
- Clean air strainers. Felt and foam air strainers should be blown clean or washed in nonexplosive solvent, allowed to dry, and reinstalled. A dirty air strainer decreases compressor efficiency and will increase oil usage.
- Clean or blow off fins on cylinders, heads, intercoolers, aftercoolers, and any other parts of the compressor or outfit that collect dust or dirt. A clean compressor runs cooler and provides longer service.
- Check the oil filter in the air line and change the filter element if necessary.

MONTHLY

- Add or change the compressor crankcase oil. Under *clean* operating conditions, the oil should be changed at the end of five hundred running hours or every six months, whichever occurs first. If operating conditions are *not clean,* change oil more frequently. SAE 10W-30, a multigrade oil, can be used as a substitute when SAE 10 or 20 weight oil is not readily available. Multigrade oils contain additives that can cause harmful carbon residue and varnish, however. Detergent-type oils are satisfactory if used before hard carbon deposits have developed. Before changing to a detergent-type oil, pistons, rings, valves, and cylinder head should be cleaned, since the detergent oil can loosen hard carbon deposits that can plug passages and damage cylinders and bearings.
- Adjust the pressure switch cut-in and cut-out settings.
- Check relief valve or CPR for exhausting head pressure each time the motor stops.
- Tighten belts to prevent slippage. A heated motor pulley is a sign of loose belts. Overtightening of belts can cause motor overload or premature failure of motor and compressor bearings.
- Check and align a loose motor pulley or compressor flywheel. It will be necessary to remove the front section of the enclosed belt guard.
- Tighten all valve plugs and covers on the compressor head to insure that a valve does not become loose and damage the valve or piston.
- Check for air leaks on the compressor outfit and air piping system.
- Check compressor pump-up time when the air receiver outlet valve is closed.
- Listen for unusual noises.
- Check and correct oil leaks.
- Perform weekly maintenance.

4.5 AIR SYSTEM SAFETY

An air compressor system is a very safe arrangement to operate. Accidents seldom happen, but the few that do occur can usually be traced to human error. To lessen the chance of human error, keep in mind the following safety precautions that should always be observed:

- **Read the instructions.** Learn what each part of the compressor does by carefully reading the owner's manual that comes with the unit.
- **Inspect before each use.** Carefully check the hoses, fittings, air control equipment, and overall appearance of the compressor before each use. Never operate a damaged unit.
- **Proper electrical outlets.** Electrical damage often results from using improperly grounded outlets. Use only a properly grounded outlet that will accept a three-prong plug.
- **Always run the compressor on a dry surface.** The compressor should be located where there is a circulation of clean, dry air. Avoid getting dust, dirt, and paint spray on the unit.
- **Starts and stops.** Most compressors start and stop automatically. Never attempt to service a unit that is connected to a power supply.
- **Keep hands away.** Fast-moving parts will cause injury. Keep fingers away from the compressor while it is running. Do not wear loose clothing that will get caught in the moving parts. Unplug the compressor before working on it.
- **Keep the belt guard on.** Use all the safety devices available and keep them in operating condition. Also, remember that compressors become hot during operation. Exercise caution before touching the unit.
- **Release air slowly.** Fast-moving air will stir dust and debris. *Be safe!* Release air slowly by using a pressure regulator to reduce pressure to that recommended for the tool.
- **Keep air hose untangled.** Keep the air, power, and extension cords away from sharp objects, chemical spills, oil spills, and wet floors. All of these can cause injury.
- **Depressurize the tank.** Be sure the pressure regulator gauge reads zero before removing the hose or changing the air tools. The quick release of high-pressure air can cause injury.

REVIEW QUESTIONS

1. Technician A says that a two-stage compressor has two different-sized cylinders. Technician B says that the more horsepower an air compressor has, the more powerful it is. Who is right?
   a. technician A
   b. technician B
   c. both technician A and technician B
   d. neither technician A nor technician B

2. Diaphragm-type compressors are used in most high-volume paint shops.
   a. True
   b. False

3. Technician A says that free air cfm is the amount of free air in cubic feet that the compressor can pump in one minute at working pressure. Technician B says that the previous statement describes displacement CFM. Who is right?
   a. technician A
   b. technician B
   c. both technician A and technician B
   d. neither technician A nor technician B

4. The displacement of a two-stage compressor is determined by the volume of the
   a. first-stage cylinder.
   b. second-stage cylinder.
   c. Both a and b are correct.
   d. Neither A nor B is correct.

5. Technician A says that fuses on the disconnect switch should be 3 1/2 times the motor current rating. Technician B says that the fuses should be 1 1/2 times the motor current rating. Who is right?
   a. technician A
   b. technician B
   c. both technician A and technician B
   d. neither technician A nor technician B

6. A 5 hp compressor that produces 15 cfm with a piping system 100 feet in length should have _____ diameter pipe.

7. Technician A says that an air lubricator should be on all air lines in a paint shop. Technician B says that an air transformer removes water from the air line. Who is right?
   a. technician A
   b. technician B
   c. both technician A and technician B
   d. neither technician A nor technician B

8. The two types of air driers are refrigeration and _____.

9. Technician A says that 1/4-inch hose should not be used in a paint shop. Technician B says that 5/16-inch hose can be used in a paint shop. Who is right?
   a. technician A
   b. technician B
   c. both technician A and technician B
   d. neither technician A nor technician B

10. In clean operating conditions, the air compressor oil should be changed every five hundred hours.
    a. True
    b. False

Chapter

5

Minor Auto Body Repairs

Objectives

After reading this chapter, you will be able to:
- List the different types of body fillers.
- Choose the correct plastic body filler for a particular repair job.
- Identify the correct way to mix filler and hardener.
- Explain how to repair dents, scratches and chips.

Key Term List

catalyst
featheredge
putty

Plastic body filler is the finishing touch to most sheet metal repairs. Restoring bent and stretched metal to its exact original shape and dimensions would be very time-consuming and almost impossible in many instances. But after the basic shape and soundness of the damaged panel has been restored (within one-eighth inch of its original contour) with proper metalworking techniques, the remaining minor blemishes can be quickly and easily covered with a thin coat of body filler. However, very careful attention must be given to the preparation and application of plastic fillers. The permanence of the repair and the quality of the final finish is adversely affected by filler that is improperly mixed and applied. While most body repairs, including minor ones, are done by the collision repair facility, the refinisher should have a knowledge of the materials and the basic procedures.

5.1 — BODY FILLERS

Most auto body repairs require some application of plastic body filler. Plastic body filler is a fast, inexpensive way to restore the final contour of a damaged panel. But many collision repair facilities tend to skimp on the sheet metal repairs and simply hide the damage under a thick layer of filler. Body fillers were never meant to replace proper metalworking techniques. Before any fillers are applied, the damaged panel should be returned to its correct shape and dimensions by bumping, picking, and pulling. Stretched metal should be shrunk, and high spots should be lowered. After the panel has been filed to locate low spots, low areas should be bumped or picked up so that no area is more than one-eighth inch below the original contour of the panel.

Before any filler is applied, all holes, cracks, and joint gaps must be welded. Conventional body fillers are very hygroscopic, which means they absorb moisture like a sponge when exposed to humid conditions.

If the holes in metal are not welded and a conventional, not water proof, type of filler is used, the filler will absorb moisture through the holes. The moisture will penetrate to the metal where rust will begin to form. Eventually, the rust will destroy the bond between the filler and the metal.

Body fillers can also be used to repair minor defects, such as dings. Procedures for making this type of minor auto body repair are given in this chapter. Be aware, though, that plastic body fillers have limitations. Large panels such as hoods, deck lids, and door panels tend to vibrate under normal road conditions. Vibrations can crack and dislodge filler that is applied over an area that is too large or applied too thickly.

Care must also be taken when applying filler to semi-structural panels in unibody frames. Panels such as quarter panels and roofs absorb road shocks and torque flexing. Excessive fillers applied in these areas can be popped off by stresses in the panels. Plastic filler should also be used sparingly on rocker panels, lower rear wheel openings, and other areas subject to flying stones and rock chips. Nor should protruding body lines, fender or door edges, or other edges and corners that are subject to scrapes and bumps be shaped with filler.

CAUTION: Plastic body fillers are hazardous materials. Always wear an approved dust mask when sanding filler. Filler dust can damage the eyes, throat, lungs, and liver.

GENERAL DESCRIPTION

Most plastic body fillers have a polyester resin that acts as a binder. The basic pigment or filler in conventional fillers is talc. Talc, also used in baby powders, absorbs moisture. That is good for the baby but bad for the car if steps are not taken to shield the filler from moisture. If holes in the metal or cracks in the paint expose the filler to the atmosphere, the talc in the filler absorbs moisture, which attacks the metal substrate and forms rust. The rust destroys the filler-to-metal bond, causing the filler to fall off. Waterproof fillers are available. Fiberglass strands or metal particles are used instead of talc as pigments.

Plastic fillers harden by chemical action. Hardening, or curing, produces a molecular structure that will not shrink or soften. The chemical reaction is set off by oxygen. If the container of plastic filler is open and left exposed to the oxygen in the atmosphere, it will slowly harden. To speed up the process, a chemical

FIGURE 5–1   Hardener catalyzes the reaction between the body filler and oxygen.

TABLE 5–1: EFFECT OF TEMPERATURE ON WORKING TIME	
Temperature	**Working Times**
100 °F	3 to 4 minutes
85 °F	4 to 5 minutes
77 °F	6 to 7 minutes
70 °F	8 to 9 minutes

catalyst is provided by the manufacturer. The catalyst, in liquid or cream form, is called *hardener* (Figure 5–1). Hardener is basically a chemical compound called peroxide. The oxygen in the peroxide drastically speeds up the curing process. As Table 5-1 shows, the filler will soon become too stiff to work in just a few minutes after adding hardener, depending on the ambient air temperature.

As the filler cures and hardens, the chemical reaction produces a tremendous amount of heat. For this reason, unused catalyzed filler should not be discarded in trash cans containing solvent-wet paper or cloths.

Curing fillers also produce a waxy coating, or paraffins, on the surface. The purpose of the paraffins in the filler is to form a film that prevents oxygen absorption from the atmosphere. The paraffins are suspended in the filler solvent and are carried to the surface when the solvents evaporate. The paraffins must

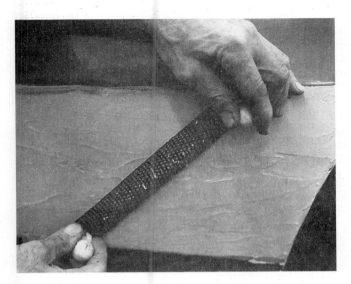

FIGURE 5–2 A cheese grater can be used to shape filler.

be either removed with a wax and grease remover before being sanded or else filed off with a surform cheese grater (Figure 5–2).

TYPES OF BODY FILLERS

During the first fifty years of auto body repair, blemishes in sheet metal panels were corrected by applying lead filler. Lead filler or solder is an alloy of lead and tin. A welding torch is used to soften the solder and bond it to the body sheet metal. Before World War II, automobiles were made with heavy-gauge steel panels that were unaffected by the heat used in the "tinning" operation. But changes began to take place in automotive construction in the late 1940s and early 1950s. In the economic boom following World War II, Americans began demanding larger and fancier cars. Manufacturers responded with vehicles made with thinner, larger, and more complex body panels. The thinner metals, however, made the old lead repair methods almost obsolete. The heat required for the lead filler warped the thin panels, and hammer-and-dolly work stretched metals too thin for filing. There was a real need for an inexpensive, time-saving substitute.

In the early 1950s, epoxy-based fillers were developed. Usually mixed with aluminum powder, epoxy fillers cured very slowly and did not harden at all if applied too thickly.

In the middle 1950s the first polyester resin-based body fillers were developed. These fillers were made from the same resin used to make fiberglass boats and required mixing with a liquid hardener and accelerator. Since the fiberglass resin is very brittle when cured and depends on cloth or matte for flexibility, the early polyester body fillers were also very brittle and hard.

The early fillers were composed of approximately 40 percent (by weight) polyester resin and 60 percent talc. Because the thin resin was difficult to work with and cured hard and brittle, many applications eventually cracked and fell off the vehicle.

Early fillers had other problems, too. Plastic filler technology had still to develop the inhibitors used today to promote stability. Therefore, the shelf life of the early fillers was very short. The original products were sold in quart cans only, yet a very high percentage of the product hardened in the can before it could be used. Early fillers also used a clear liquid hardener that made it difficult to determine when the filler and hardener were thoroughly mixed. Incomplete mixing often resulted in soft spots in the repair area.

When the filler and hardener were properly mixed together, the filler dried very hard. Early fillers were difficult to file and had to be leveled with a grinder, resulting in choking clouds of dust that blanketed the shop. Low dust, straight-line air files had not been developed yet.

Finally, a product was developed that utilized more flexible polyester resins and benzoyl peroxide as a cream hardener. Black pigment was added to the resin and talc mixture, and white pigment was added to the cream hardener. The contrasting colors provided a reference to ensure proper mixing of the two ingredients.

As the technology developed, body fillers became softer, easier to apply, and easier to shape. Fillers soon appeared in black, red, gray, white and yellow. Cream hardeners in contrasting colors—red, white, green, and blue—were also developed to provide a mixing reference for the various colored polyester fillers. The softer fillers could also be grated while still semi-cured, thus reducing the amount of sanding required. Note that the addition of color does not affect the working characteristics of the filler.

Conventional body fillers have over thirty years of development backing them today. The premium heavyweights use very fine grain talc to provide superior workability, sandability, and featheredging. High-quality resins ensure excellent adhesion and quick curing properties. Most heavyweights can be grated in ten to fifteen minutes.

Table 5-2 summarizes the ingredients, characteristics, and applications of body fillers and putties.

TABLE 5–2:   COMPARING FILLERS

Filler	Composition	Charateristics	Application
Conventional Fillers			
Heavyweight Fillers	Polyester resins and talc particles	Smooth sanding; fine featheredging; nonsagging; less pinholing than lightweight fillers	Dents, dings, and gouges in metal panels
Lightweight Fillers	Microsphere glass bubbles; fine-grain talc; polyester resins	Spreads easily; nonshrinking homogenous; no setting	Dings, dents, and gouges in metal panels
Premium Fillers	Microspheres; talc; polyester resins; special chemical additives	Sands fast and easy; spreads creamy and moist; spreads smooth without pinholes; dries tack-free; will not sag	Dings, dents, and gouges in metal panels
Fiberglass-Reinforced Fillers			
Short Strand	Small fiberglass strands; polyester resins	Waterproof; stronger than regular fillers	Fills small rustouts and holes. Used with fiberglass cloth to bridge larger rustouts.
Long Strand	Long fiberglass strands; polyester resins	Waterproof; stronger than short-strand fiberglass fillers; bridges small holes without matte or cloth	Cracked or shattered fiberglass. Repairing rustouts, holes, and tears.
Specialty Fillers			
Aluminum Filler	Aluminum flakes and powders; polyester resins	Waterproof; spreads smoothly; high level of quality and durability	Restoring classic and exotic vehicles
Finishing Filler/Polyester Putty	High resin content; fine talc particles; microsphere glass bubbles	Ultra-smooth and creamy; tack-free; nonshrinking; eliminates need for air dry–type glazing putty	Fills pinholes and sand scratches in metal, filler, fiberglass, and old finishes.
Sprayable Filler/Polyester Primer-Surfacer	High-viscosity polyester resins; talc particles; liquid hardener	Virtually nonshrinking; prevents bleed-through; eliminates primer/glazing/primer procedure	Fills file marks, sand scratches, mildly cracked or crazed pint films, and pinholes. Seals fillers and old finishes against bleed-through.

Fiberglass Fillers

As thinner-gauge sheet metal replaced the heavy-gauge steel used on vehicles of the 1940s and 1950s, rust became a problem, especially in areas of the country where road salts were used in winter. A product was needed to repair rustouts. Because talc-filled body fillers absorb moisture readily, the available heavyweight fillers did not provide long-lasting protection when used to repair rustouts.

To meet this demand for a waterproof filler, fiberglass-reinforced fillers were developed. Fiberglass fillers use fiberglass strands rather than talc as a bulking agent. These fillers are more flexible and stronger than conventional fillers. Because they are also waterproof, they can be used to bridge holes, tears, and rustouts.

Fiberglass fillers are available in two basic forms. One is formulated with short strands of fiberglass. The other is made with long strands. Short-strand fiberglass fillers are generally used to repair small holes (approximately 1 to 1-1/2 inches in size). When used to repair larger holes, a fiberglass cloth or screen should be used as a back support. The short-strand fillers can be sanded and finished like any conventional filler.

Long-strand fiberglass products are designed to fill holes larger than 1-1/2 inches in size. The longer strands interlock and provide a much stronger patch. The long-strand filler might also be used with fiberglass cloths or mats to bridge even larger rustouts. The long-strand fillers, however, are used only as a base. Smoother fillers, either short-strand fiberglass fillers or conventional fillers, must be used for the final fill. Chopped fiberglass fibers are also available to be added to fillers to increase their strength.

Aluminum Fillers

Some manufacturers tried to improve the water resistance of their products by replacing part of the talc with aluminum powder. This small quantity of aluminum powder did not stop moisture failures because of the talc still remaining in the formula. Another complaint was the short shelf life of the aluminized products, since aluminum is itself a catalyst for polyester resin.

The first talc-free, 100 percent aluminum auto body filler was developed in England and was introduced into the United States under the name Alum A Lead. The shelf life problem was solved by packaging the resin and powder separately; mixing was done by the refinisher.

The first premixed, 100 percent aluminum auto body fillers were introduced in 1965. This product was waterproof, used a red-tinted liquid hardener, and had a fairly good shelf life. Due to their very high relative costs, the 100 percent aluminum-filled body fillers are used sparingly on special applications, such as restoring antique cars. Today there are several similar 100 percent aluminum products available. Metal fillers are nonshrinking, waterproof, and very smooth. When cured, they are harder than talc or fiberglass-filled plastic fillers.

Lightweight Fillers

Until the middle 1970s only minor improvements were made to conventional heavyweight fillers. With the invention of microsphere glass bubbles by 3M came the technology to produce the modern, lightweight auto body fillers of the 1980s (Figure 5–3). Lightweight fillers were formulated by replacing about 50 percent of the talc in the filler with tiny glass spheres. The resulting higher resin content dramatically improved the

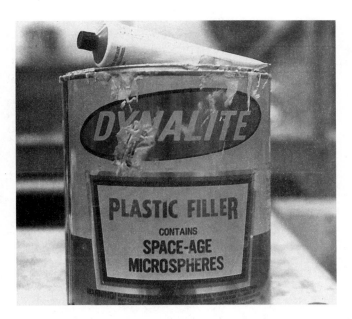

FIGURE 5–3 *Lightweight fillers contain microsphere glass beads.*

filing and sanding characteristics of the filler as well as improved the filler's adhesion and water resistance. Most lightweight fillers are homogenous. The glass bubbles remain suspended in the resin and do not settle to the bottom of the can. This homogenous composition allows lightweight fillers to be packaged in plastic bags or cans and dispensed with rollers or compressed air or squeezed out with a plastic spreader (Figure 5–4). The plastic bags keep the filler fresh and eliminate much of the wasted filler sometimes associated with canned fillers.

By the middle 1980s lightweight body filler technology was shared by most manufacturers, and major brands again became very similar to each other in working characteristics. Lightweight fillers quickly became the most popular filler used. Nationally, lightweight fillers represent more than 80 percent of the total filler in paint shops.

Premium Fillers

Filler manufacturers in the mid-1980s took advantage of new technology to produce premium-quality fillers. Premium fillers have superior performance qualities that go beyond the capabilities of conventional lightweight fillers. Premium fillers are moist and creamy. They spread easily yet will not sag on vertical surfaces. They dry tack-free without pinholing. Best of all, premium fillers are easy to sand. The smooth finish and ease of sandability reduce the time and labor involved in filling and shaping the repair.

FIGURE 5–4 A five-gallon pneumatic can dispenser.

FIGURE 5–5 Polyester glazing putty.

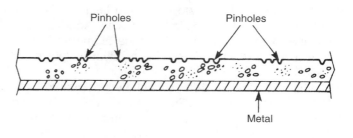

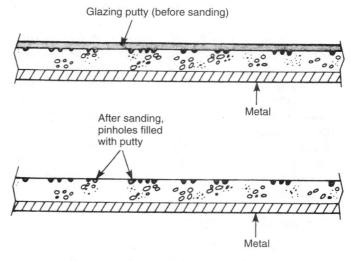

FIGURE 5–6 Pinholes and scratches in the body filler can be filled with polyester glazing putty.

Polyester Glazing Putty or Two-Part Spot Putty

In the 1980s, the European basecoat/clear coat paint system became popular in the United States. The new basecoat/clear coat paint systems stirred up a problem that had occurred occasionally since the inception of the candy and mother-of-pearl colors in the late 1960s. The rich solvents and multicoats required for these "trick" paint jobs caused the pigment from the cream hardener in the body filler to bleed and stain the finish of light colors, usually after several days of exposure to sunlight. The widespread use of basecoat/clear coat products and other multicoat systems has made this staining problem a more frequent occurrence.

Developing a body filler that will not stain results in either extremely high cost or working properties not acceptable to most body technicians. At present, the only absolutely stain-free fillers are liquid hardener-catalyzed aluminum fillers.

To solve the staining problem, body filler manufacturers have developed a fine-grained, catalyzed polyester glazing **putty** (Figure 5–5). Polyester glazing putty does not shrink, has excellent dimensional stability, and resists solvent penetration (the cause of bleed-through). When applied over traditional body fillers, polyester glazing putties effectively solve the bleed-through problem. Mixing, applying and shaping of body filler usually creates tiny pinholes and sand scratches (Figure 5–6). Polyester glazing putty can also be used to fill these minor imperfections. Use a

FIGURE 5–7 *Apply polyester glazing putty with a razor blade.*

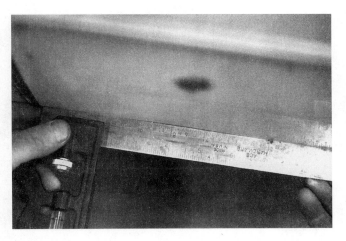

FIGURE 5–8 *A straightedge will rock on the high spots and gap on the low spots. This example is a gap.*

razor blade to apply the catalyzed putty (Figure 5–7). The razor's edge applies the minimum amount of putty to fill the hole. More than the minimum amount of putty will require extra sanding time. Allow the putty to cure. Then lightly sand with 180-grit paper on a block.

5.2 — DOOR DING REPAIR

Dings are simple to repair if you eliminate high areas and fill low areas. First wash the surface with soap and water to remove water-soluble contaminants. Next clean with a wax and grease remover. Locate high areas with a straightedge (Figure 5–8). If the straightedge rocks on a point, you have located a high spot. Gaps between the straightedge and the panel indicate lows.

Eliminate high spots by metal filing or grinding. If you use a grinder, start with a fresh 24-grit disk. Usually a high spot surrounds the dent. Hold the grinder so that only the upper one inch of the disc is in contact with the metal. Quickly grind the high spot, moving from top to bottom, then grind again, moving from bottom to top. Pass over the high spot quickly to prevent excessive heat buildup. If this does not eliminate the high spot, tap with a body hammer. Check with the straightedge to make sure the high spot is gone. Carefully grind the paint away from the fill area (Figure 5–9).

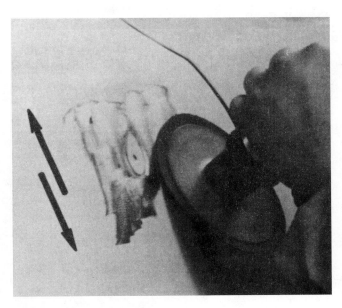

FIGURE 5–9 *Grinding off the high spot surrounding a ding. Remove all of the paint from the fill area.*

Mix the plastic filler and hardener according to the directions on the can. Apply the filler to the low area with a spreader. Press the filler down into the dent with the spreader. This will insure that the filler will adhere. A slight overfill should be applied. Filler may be applied over bare metal or bare metal covered with one coat of cured epoxy primer. Because of adhesion problems, filler should not be applied over paint. Allow the filler to cure. Test for cure by scratching. If the filler has not cured, it will not adhere where it is sanded; the edges will roll off. Begin sanding with a

FIGURE 5–10   On a high-crown surface, keep the file on edge.

FIGURE 5–11   Hold the air file flat when sanding a low-crown panel.

flat board. On high crown surfaces, hold the flat board on edge (Figure 5–10). This will allow the board to follow panel contour. On low crown panels, the board should be held flat (Figure 5–11). Use the undamaged metal surrounding the filled dent as a guide. To do this, the board must bridge the filler so that the ends of the board are supported by undamaged metal.

Once the filler has been sanded to level with 40-grit paper, a coat of two-part spot putty can be applied. The putty is mixed and applied in the same way as filler. The putty is allowed to cure, then it is sanded with 80-grit sandpaper followed with 180-grit sandpaper. If two-part putty is not used, the filler is simply sanded with 80-grit sandpaper followed with 180-grit

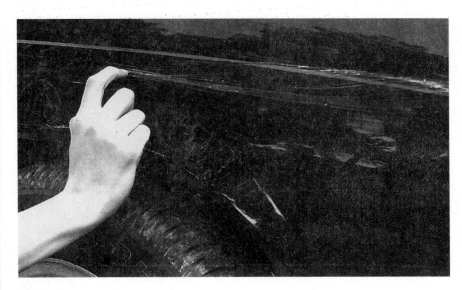

FIGURE 5–12 This Mustang was scraped by a pickup truck. The technician uses a fingernail to determine the depth of the scratches.

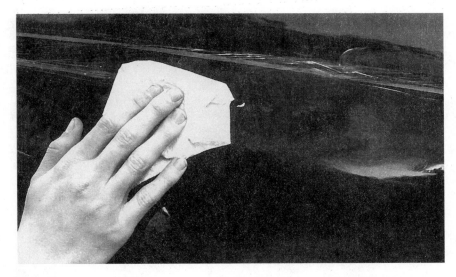

FIGURE 5–13 The technician uses rubbing compound to remove the superficial scratches.

sandpaper. The repair is now ready for featheredge (see Chapter 9).

5.3 — SCRATCH REPAIR

Drag your fingernail across the scratch (Figure 5–12). If your fingernail does not catch on the scratch, the scratch can be buffed out. If your fingernail does catch on the scratch, sanding with 1500-grit paper and buffing may remove it (Figure 5–13). Check the

paint thickness with a mil gauge. There should be at least 4 mils of paint. Chapter 15 discusses how to buff. If buffing does not eliminate the scratch, follow the procedure for chips.

5.4 — CHIP REPAIR

To remove a single chip, first wash the surface with soap and water. Follow this step with a wax and grease remover. In order to keep surface prep to a

FIGURE 5–14   A flat stick wrapped with sandpaper can be used to featheredge chips. A DA held flat could also be used.

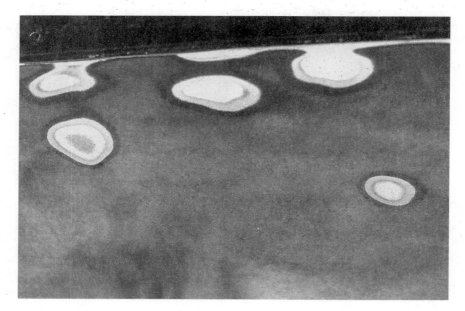

FIGURE 5–15   Featheredged chips in a hood.

minimum area, use a small one- to three-inch flat board to sand (Figure 5–14). This can be made by simply cutting a wooden ruler or flat paint stick to the desired length. Wrap the flat board with 320-grit paper. Sand on the chip to featheredge the paint layers (Figure 5–15). On a high-crown surface, you can hold the board on edge to prevent creating a flat spot. When the area is featheredged, it can be primed, surfaced, and guide coated.

If there are many paint chips, such as the front edge of a hood, it is best to remove all of the paint. Use a DA with 80-grit paper. It will rapidly remove the paint.

REVIEW QUESTIONS

1. Technician A says that at cooler temperatures, body filler cures faster. Technician B says that at warmer temperatures, body filler cures faster. Who is right?
   a. technician A
   b. technician B
   c. both technician A and technician B
   d. neither technician A nor technician B
2. Fiberglass fillers are waterproof.
   a. True
   b. False
3. Technician A says that cured aluminum filler is harder than fiberglass filler. Technician B says that cured aluminum filler is waterproof. Who is right?
   a. technician A
   b. technician B
   c. both technician A and technician B
   d. neither technician A nor technician B
4. Microsphere glass bubbles are used in _____ filler.
5. Technician A says that premium fillers are used to fill pinholes. Technician B says that premium fillers are easy to sand. Who is right?
   a. technician A
   b. technician B
   c. both technician A and technician B
   d. neither technician A nor technician B
6. High spots can be removed in ding repair by
   a. grinding.
   b. filing.
   c. heat.
   d. a and b only.
7. Technician A uses a flat stick to feather edge a repair area. Technician B applies plastic filler over sanded paint. Who is right?
   a. technician A
   b. technician B
   c. both technician A and technician B
   d. neither technician A nor technician B
8. Cured body filler is first sanded with _____-grit paper.
9. Technician A uses a spreader to apply glazing putty to pinholes. Technician B uses a razor blade to apply glazing putty to pinholes. Who is right?
   a. technician A
   b. technician B
   c. both technician A and technician B
   d. neither technician A nor technician B
10. Body filler should be applied to at least one-quarter inch of thickness.
   a. True
   b. False

Chapter

6

Restoring Corrosion Protection

Objectives

After reading this chapter, you will be able to:
- **Define corrosion and describe the common factors involved in rust formation.**
- **Describe the anticorrosive materials used to prevent and retard rust formation.**
- **Explain the conditions and events that lead to corrosion on the auto body.**
- **Choose the correct anticorrosive application equipment for specific applications.**
- **Outline the correct corrosion treatment procedures for each of the four general corrosion treatment areas.**
- **List the four types of seam sealers and explain where each should be used.**

Key Term List

corrosion
rustproofing
galvanizing
pH
epoxy primer
self-etching primer
seam sealer
acid rain

Corrosion is a problem that has always concerned paint shop technicians and refinishers. It requires either repair work (see Chapter 5) or special treatment when refinishing (see Chapter 9). But, with the following recent developments in the automotive industry, corrosion prevention has taken on new meaning to paint shop personnel:

- Once an auto body has been repaired after a collision, the rustproofing, undercoating, and sound deadening should be restored.
- Car manufacturers are including in their owner's manuals recommendations for sheet metal repair

or replacement. They suggest that the paint shop should apply an anticorrosive material to the part repaired or replaced so that corrosion protection is restored.
- In affiliation with the Inter-Industry Conference on Auto Collision Repair (I-CAR), the insurance companies are promoting corrosion prevention repair to the paint shops.
- The increased usage of replacement panels in the paint shops requires widespread corrosion prevention treatment.
- Possibly the major reason, however, is the advent of the unibody car.

In unibody construction, the car's body panels are no longer cosmetic sheet metal. They now constitute the structural integrity of the vehicle. This means that rust is not just an eyesore. The unibody car has more welded joints in critical structural areas where corrosion can do serious damage. It is an ever-present danger to the unibody vehicle, since rusting of structural panels and rails can affect the driveability of the car and the safety of its passengers.

6.1 WHAT IS CORROSION?

Corrosion—or rust, when it occurs on steel—is the product of a complex chemical reaction with serious, costly consequences (Figure 6–1). Chemical corrosion requires three elements (Figure 6–2):

- Exposed metal
- Oxygen
- Moisture (electrolyte)

In other words, the formula for rust in a car body is:

Iron + Oxygen + Electrolyte = Rust or Iron Oxide

Three basic types of corrosion protection used on today's automobiles are:

- Galvanizing or zinc coating
- Paint
- Anticorrosion compounds

FIGURE 6–1 Closeup of car's number one enemy—rust.

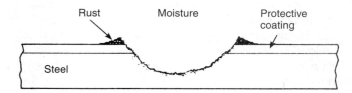

FIGURE 6–2 Breakdown in protective coating causes rapid rust formation.

Galvanizing is a process of coating steel with zinc. It is one of the principal methods of corrosion protection applied during the manufacturing process. On galvanized steel, the zinc forms a natural barrier between the steel and the atmosphere. As the zinc corrodes, a layer of zinc oxide will form on the surface exposed to the atmosphere. Unlike iron oxide or rust, the zinc oxide adheres to the zinc coating tightly, forming a natural barrier between the zinc and the atmosphere. When the surface of the car's finish is damaged by a scratch or nick, the zinc coating undergoes corrosion, sacrificing itself to protect the iron under it. The resulting zinc oxide actually forms a protective coating and repairs the exposed area of the steel. Thus, zinc performs a twofold protective process: first, it provides a natural barrier, and second, if it is exposed to the atmosphere, it is transformed into a protective zinc oxide coating over the exposed steel.

A paint system such as those described in later chapters of this book will provide a barrier between the atmosphere and the steel surface. When this barrier is in place (Figure 6–3), the moisture and impurities in the air cannot interact with the steel surface, and the steel is protected from corrosion. If the paint surface or barrier is broken by a stone chip or scratch, the steel in this area is no longer isolated from the moisture and impurities in the air. Corrosion will then take place in this region. Corrosion will spread between the paint and steel surface. If the adhesion of the paint to the steel is poor, large sections of the paint can be separated from the steel. This will result

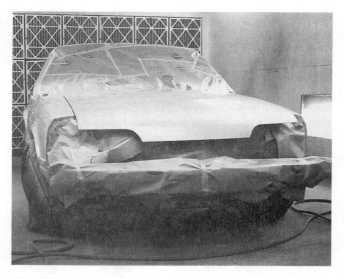

FIGURE 6–3 New paint restores the corrosion protection to the outside of this vehicle.

in a large area of the steel being left unprotected, and severe rust in this region will quickly follow. If impurities are present between the paint and the steel, oxygen in the air can pass through the paint, reacting with the impurities and the steel to form rust. In this case, corrosion will take place on the steel surface and the protective paint barrier will be destroyed. Paint, by itself, is only effective as long as the paint film remains intact.

Anticorrosion compounds are additional coatings applied over the paint film. Protective coatings can be applied either by the manufacturer or as an aftermarket process. The two most popular types of anticorrosion coatings are:

- Petroleum-based compounds
- Wax-based compounds

Anticorrosion compounds are primarily used in enclosed body sections (Figure 6–4) and other rust-prone areas.

The auto manufacturers have increased their corrosion protection measures. New processes and methods, including the use of coated steels, zinc-rich primers, and more durable base coatings have made it possible for modern cars to survive corrosive forces for longer periods than before. The following is a typical new car finishing sequence (Figure 6–5) used by major auto manufacturers:

1. Use coated or galvanized steel (Figure 6–6).
2. Chemically clean and rinse.
3. Apply conversion coating.
4. Apply **epoxy primer**.
5. Bake primer.
6. Apply surfacer.
7. Apply color coats.
8. Bake color coats.
9. Apply anticorrosion materials.

Because of these better finishing procedures, corrosion protection warranties (Figure 6–7) of up to ten years are now a reality. With these dramatic improvements in the performance of OEM products, the repair industry must rise to the challenge of producing corrosion resistance in repaired areas that matches or exceeds the durability of the original product. Repair work that does not stand up will draw attention to itself next to the outstanding durability of many original finishes. It can also draw liability challenges where issues of vehicle safety are involved. Remember, the paint shop technician is responsible for the quality and durability of the repairs completed. Remember that the customer is entitled to a car restored to the way it was before the damage occurred.

6.2 — CAUSE FOR THE LOSS OF FACTORY PROTECTION

Even with all of the care taken to protect vehicles, breakdown still occurs. The breakdown of corrosion protection falls into three general categories:

- Paint film failure
- Collision damage
- Repair process

The paint film is the result of the entire process of coatings, primers, and color coats that the manufacturer applied. When the paint film fails, corrosion begins. Stone chips (Figure 6–8), moisture, and improper surface preparation can all lead to film failure.

During a collision, the protective coatings present on a car are damaged (Figure 6–9). This occurs not just in the areas of direct impact, but also in the

FIGURE 6–4 Anticorrosion compound sprayed inside a door.

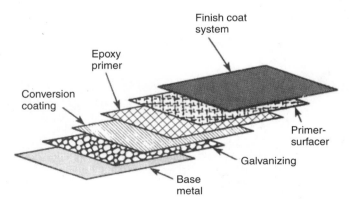

FIGURE 6–5 Typical buildup of corrosion prevention material used by car manufacturers.

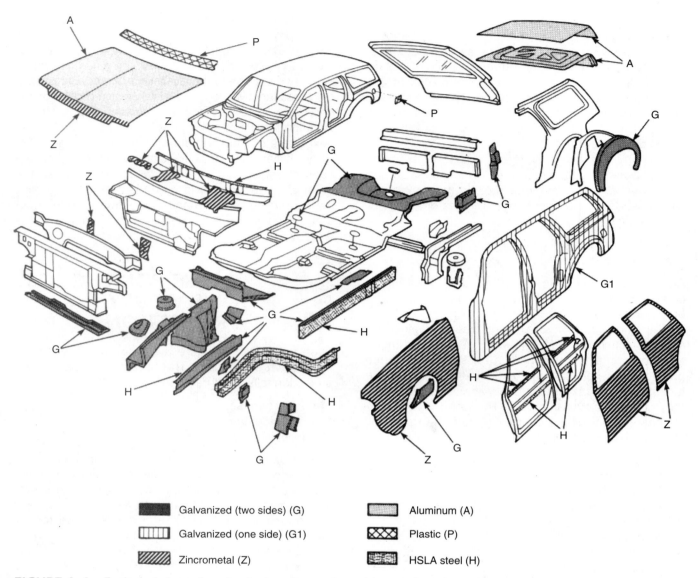

Galvanized (two sides) (G)

Galvanized (one side) (G1)

Zincrometal (Z)

Aluminum (A)

Plastic (P)

HSLA steel (H)

FIGURE 6–6   Exploded view of car body showing parts and types of coating.

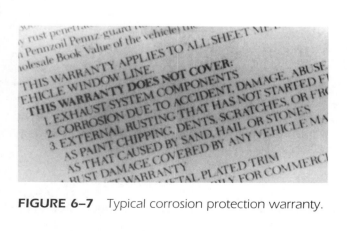

FIGURE 6–7   Typical corrosion protection warranty.

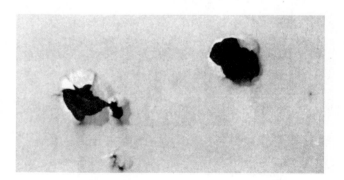

FIGURE 6–8   Stone chips can lead to rust spots.

FIGURE 6–9 When this vehicle is repaired, the corrosion protection must be restored.

FIGURE 6–11 Clamping used to hold vehicles during pulling can remove paint. All scraped areas must be treated.

FIGURE 6–10 Heat from welding and cutting operations destroys factory corrosion protection.

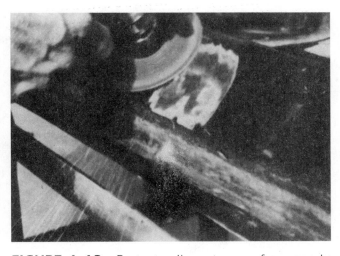

FIGURE 6–12 Protect adjacent areas from sparks when grinding or welding.

indirect damage zones. Seams pull apart, caulking breaks loose, and paint chips and flakes. Locating and restoring the protection to all affected areas remains a key challenge for the paint technician.

Vehicle repair is possibly one of the major causes of protective coating damage. For example, repair procedures often require cutting body panels and seams either mechanically or with a plasma torch. Even minor straightening and stress-relieving procedures can damage these protective coatings so corrosion can start. Normal welding temperatures cause zinc to vaporize and be lost from the weld area (Figure 6–10). Abrasive operations during repair and refinishing can also leave areas unprotected. After all welding and repair work has been completed, these damage points need careful attention to eliminate contaminants. Then steps must be taken to exclude

the atmosphere from the metal by sealing all surfaces thoroughly.

Other precautions that should be taken to protect the factory corrosion protection are:

- Remove only the minimum amount of paint film from affected areas such as welded points.
- Be extremely careful not to scratch any part except that to be repaired.
- When clamping or holding the affected panels during body repair work (Figure 6–11), clamping tools can cause scratches on the panel. They must be treated to avoid rusting.
- While grinding (Figure 6–12), cutting, or welding panels, place protective covers over adjacent painted surfaces and surrounding areas to protect them from the flame or metal chips.

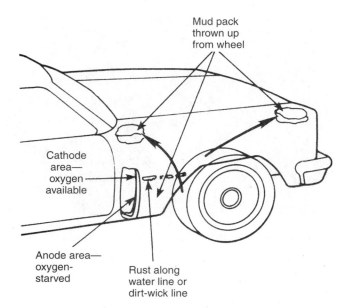

Drain

FIGURE 6–13 *Inside-out corrosion caused by mud and debris packed against the underside of fenders is usually undetected until it eats through the metal.*

FIGURE 6–14 *Door drain hole.*

- Cover any opening of the body sills and similar area with masking tape to prevent metal chips from entering during the grinding, cutting, or welding operation.
- Completely remove any metal chips from inside the body. Use a vacuum cleaner, not dry compressed air, to remove metal chips. If dry compressed air is used, metal chips can be blown out and accumulate in corner areas.

There are also some environmental and atmospheric conditions that help to influence the rate of corrosion. They are:

- **Moisture.** As the amount of sand, dirt, mud, and water on the underside of the body increases, so will the chances of corrosion accelerate (Figure 6–13). Floor sections that have snow and ice trapped under the floor matting will not dry. Likewise, if holes at the bottom of the doors and side sills (Figure 6–14) are not kept open, water will accumulate. Remember, water is one of the requirements for rust.
- **Relative Humidity.** Corrosion will be accelerated in areas of high relative humidity, especially those areas where the temperatures stay above freezing and where atmospheric pollution exists and road salt is used.
- **Temperature.** A temperature increase will accelerate the rate of corrosion to those parts that are not well ventilated.

TABLE 6–1: RELATIVE ACTIVITY OF METAL	
Magnesium	Most Active
Aluminum	
Zinc	
Chromium	
Iron	
Cadmium	
Cobalt	
Nickel	
Tin	
Lead	
Copper	Least Active

- **Air Pollution.** Industrial pollution and acid rain, the presence of salt in the air in coastal areas, or the use of heavy road salt will accelerate the corrosion process. Road salt will also accelerate the disintegration of paint surfaces.

Another type of corrosion that must be considered when working on automobiles is known as galvanic corrosion. This occurs when two dissimilar metals are placed in contact with each other. The more chemically active of two metals will corrode, protecting the other metal in the process. As shown in Table 6-1, this is why zinc will sacrifice itself to protect steel. This

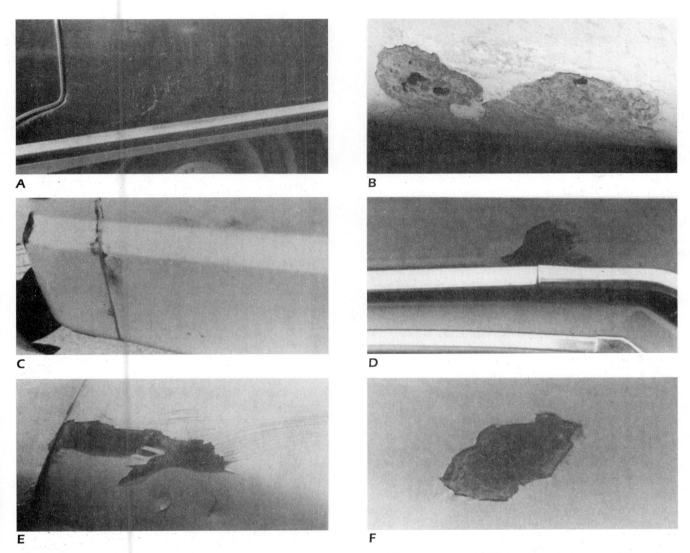

FIGURE 6–15 _Common corrosion areas: (A) wheel housing, (B) rocker panel, (C) body panel seam, (D) roof drip rail, (E) collision damage, (F) paint chip._

sacrificial corrosion of zinc prevents corrosion or rust on galvanized steel. In the case of other metals, such as aluminum and steel, galvanic corrosion can cause problems.

Regardless of the cause, if corrosion prevention is not practiced, the cost to the paint shop is comebacks or lost customers. Inadequate preparation that leaves dirt, grease, or acids on the metal will cause the loss of adhesion. Rust will start, a little at first, creating corrosive "hot spots" at the points of failure. Surface failure will progress quickly in unseen or enclosed areas, spreading under the surface coatings, eating deeper and deeper into the metal. Figure 6–15 illustrates some of the more common hot spots found on an automobile.

The main areas of concern for renewing corrosion protection generally are

- **Enclosed interior surfaces.** Includes body rails and rocker assemblies.
- **Exposed interior surfaces.** Including floor pan, apron, and hood sections.
- **Exposed joints.** Such as quarter-to-wheel-housing and quarter-to-trunk floor joints.
- **Exposed exterior surfaces.** Such as fenders, quarter panels, and door skins.

The paint shop's main interest is in the finish of the vehicle's surface. But as with minor repair techniques, the refinisher should know about the materials used in corrosion protection.

ANTICORROSION MATERIALS

The paint shop's efforts in protecting car bodies from rusting should focus on creating a clean, chemically neutral surface on the sheet metal, then sealing the material under layers of paint. Under certain conditions, as mentioned earlier in the chapter, a wax- or petroleum-based anticorrosion compound is used to exclude air and moisture from the metal surface.

More and more new vehicles come off the assembly line today with anticorrosive materials that are available to the paint shop. Being able to replace or install these wax- or petroleum-based materials is very important to the knowledgeable refinisher.

Corrosion prevention has not always been a common paint shop operation. The original rustproofing was called undercoating, and it was an asphalt-based product that was sheer agony to apply, because it got not only on the underside of the car, but also on everything else within 20 feet of the application bay. Worst of all, it did not work very well. In time, the solvents used would evaporate, the asphalt would harden and crack, and the moisture that causes oxidation would actually become trapped under the undercoating.

Asphaltic undercoating did have benefits in terms of sound deadening and preventing stone marks under fenders. And it is useful today on fiberglass panels for the same reasons. As a rustproofer, however, it probably was not the best.

When selecting a modern anticorrosive material, there are several things that should be considered:

- The material should be thin enough to flow or penetrate pinch weld cracks and to creep adequately to protect the exposed metal of such areas as the steel immediately adjacent to spot welds—a particularly tough **rustproofing** proposition.
- The material should have good adherence to both bare metal and painted surfaces. In addition to adhering to the surface, it should be highly resistant to water, to cutting from stones thrown up from the road, to ordinary solvent-type materials used in the engine and elsewhere, and so forth. In other words, it should not only protect initially, it should also continue to protect. Material that does not retain some pliability and toughness will not do the job.
- It is important to choose a material without solvents that might have a lingering bad odor that could be present in a car when it is delivered.
- The product should be easy to clean up with ordinary and safe solvents.

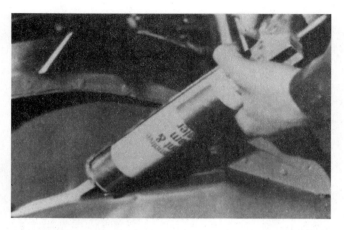

FIGURE 6–16   *Seam sealer can be applied with a caulking gun.*

Anticorrosive materials or agents can be divided into three broad categories:

- **Anticorrosion Compounds.** As already mentioned, either wax- or petroleum-based compounds are resistant to chipping and abrasion; they can undercoat, sound deaden, and completely seal the surface of a car from the destructive causes of rust and corrosion. They should be applied to the undercarriage and inside body panels so that they can penetrate into joints and body crevices to form a pliable, protective film.
- **Body Sealers or Sealants.** These prevent the penetration of water or mud into panel joints and serve the important role of preventing rust from forming between adjoining surfaces (Figure 6–16).
- **Primer.** These undercoats are applied to bare metal to promote adhesion of the top coat. There are four types of primer:

  - **Weld-Through Primer (Figure 6–17).** To properly weld two panels together, all contact areas must be bare metal. After a lap joint is welded, the lapped area cannot be primed. To get around this problem, on nongalvanized steel, the underside of the top panel and upper side of the bottom panel are sprayed with zinc-rich weld-through primer before they are fitted together. This protects the inaccessible area. Due to poor adhesion, remove any accessible weld-through primer before priming.
  - **Self-Etching Primer.** A self-etching primer contains acid to bite into the bare metal surface. It provides an excellent adhesion to the bare metal and prevents rust. Two coats of self-etching primer are recommended; any more than two coats could be harmful.

FIGURE 6–17 *Weld-through primer in use.*

- **Epoxy Primer.** Epoxy primer also provides excellent adhesion to bare metal. This adhesion prevents rust. Epoxy primer is recommended for use inside spliced frame rails.
- **Rust converter.** Rust converters change the reactive form of iron oxide (red) to a more stable form of iron oxide (black). They are water-based and can be used as sealers if a lifting problem is encountered.

SHOP TALK

Be sure to carefully read the manufacturer's instructions on the container and follow them. Several of these anticorrosive materials can be used on the same part or section.

6.4 **ANTICORROSION APPLICATION**

This section will explain how to make corrosion protection repairs to enclosed frame rails and interior and exterior joints. Surface preparation of exterior rust, welds, and bare metal is covered in Chapter 9.

The materials used in corrosion protection restoration are, in many cases, the same as those used in surface preparation. Always follow the manufacturer's safety recommendations for respirator, gloves, suit, and glasses.

When a unibody frame rail is spliced, the inside or enclosed weld area must be primed and painted to prevent rust. The weld area is a corrosion hot spot that will rapidly rust if not treated. The following steps will insure a proper repair:

1. Remove weld scale or blistered paint. If the weld is near enough to the end of the rail to reach, simply use an abrasive pad to remove loose paint. If the weld is not accessible by hand, try a long-handled bristle brush to reach the area. If this is not possible, use a blowgun to direct a stream of air into the rail to knock off the scale.
2. If the weld is accessible by hand, use a wax and grease remover to dissolve any oil-based contaminant.
3. Mix two-part epoxy primer. On the accessible rail, use a spray gun to apply two coats. Allow proper flash time between coats. On an inaccessible rail, use a cup and spray wand assembly. Insert the wand to beyond the weld (Figure 6–18). Trigger the wand and slowly pull it out. Release the trigger. Apply another coat, observing the flash time. Do not use self-etching primer because it is difficult to monitor how much self-etching primer is applied inside the rail. Too much self-etching primer could cause corrosion.
4. After the epoxy primer has set up, spray a coat of color, using a gun or wand.
5. After an overnight dry time, spray a wax-based anticorrosion compound over the paint.

FIGURE 6–18 Epoxy primer sprayed inside a frame rail.

FIGURE 6–19 This technician is using a cut-off wheel to dress welds on a pickup truck's bedside.

JOINTS

Some examples of joints are trunk floor and inner wheelhouse, inner rocker panel and floor, and apron and radiator support. The following materials are used to prevent corrosion in the repaired area:

- **Primer.** To bond to bare metal and prevent rust
- **Surfacer.** To bond to primer and fill scratches
- **Seam sealer.** To fill gaps and keep contaminant out
- **Topcoat.** To complete the repair

Follow these steps to properly treat joints:

1. Begin the repair by cleaning the area to be refinished with wax and grease remover. Use a spray bottle to apply the liquid. Wipe dry with a clean cloth.
2. On panels that need a smooth appearance, dress down the weld nuggets with a cut-off wheel (Figure 6–19). Do not remove any metal from around the weld.
3. Use a nylon wheel to clean weld scale off the welds and to remove blistered paint (Figure 6–20).

FIGURE 6–20 *An abrasive wheel can be used to remove blistered paint.*

4. Featheredge the broken paint with 80-grit paper and 320-grit paper on a DA. If a DA will not fit into the area, use a hand block or a flat wooden stick.

5. Blow out the dust and wipe with a mild wax and grease remover.

6. Mask adjacent areas to prevent overspray contamination.

7. Mix and spray two coats of self-etching or epoxy primer. Follow the manufacturer's directions for flash times.

8. If the surface must have a smooth finish, apply surfacer to fill in scratches. Allow the surfacer to cure. Refer to the paint manufacturers' recommendations for the grits of sandpaper to use when sanding the surfacer. If the area is to be blended with a clear coat, sand all areas to be clear blended with 1000-grit paper or a light abrasive pad. If the surface does not need to be smooth and some scratches are allowed, you can skip this step.

9. Apply a seam sealer. There are four types of seam sealers:

• Thin-bodied: Use on one-eighth-inch or smaller seams, spread smooth.

• Heavy-bodied: Use on one-eighth to one-quarter-inch seams, spread smooth.

• Brushable: Up to one-quarter-inch seams, brushed rough.

• Solid: One-quarter-to one-half-inch seams, pressed into place.

FIGURE 6–21 *Brushable seam sealer.*

The best choice for interior joints is brushable seam sealer, because most interior joints are covered with carpets or liners (Figure 6–21). Since most interior joints are hidden, they do not need a smooth surface.

10. Exterior joints often have a factory seam sealer. To duplicate the built-up appearance, apply masking tape above and below the joint (Figure 6–22). The distance between the two pieces should be the same as the desired width of the seam sealer. Cut off the tip of the heavy-bodied sealer tube to be the same as the desired width. Using a caulk gun, carefully

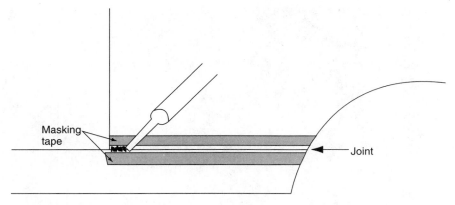

FIGURE 6–22   *Setup to duplicate factory sealer on joint.*

apply the heavy-bodied seam sealer to one edge of the joint. Use even pressure on the caulk gun trigger to make a smooth seam. Drag the gun to the other end of the joint. Remove the tape. If a built-up appearance is not required, apply the seam sealer and smooth with a paint stir stick. If the seam sealer requires priming before the topcoat, apply the primer after it sets up.

11. Apply the topcoat. Clear coat if needed.

6.5 ACID RAIN DAMAGE

As mentioned earlier, air pollutants can damage an automotive finish. Since most of their damage is done to exterior, finished surfaces, they are a major concern of the refinisher.

Acid rain and other pollutants have generated a lot of controversy in recent years, and there has been some confusion as to their causes and effects. Sulfur dioxide or nitrogen oxides create acid rain when they are released into the atmosphere and combine with water and the ozone to create either sulfuric or nitric acid. It is estimated that the United States alone pumps out thirty million tons of sulfur dioxide and twenty-five million tons of nitrogen oxides yearly. More than two-thirds of the sulfur is emitted from power plants burning coal, oil, or gas. Iron and copper smelters, automobile exhaust, and natural sources like volcanoes, wetlands, and forest fires account for most of the remaining pollutants.

The standard for measuring acid rain is the **pH** scale. The pH scale ranges from zero to 14, with 7 being neutral, neither acid nor base (the pH of pure water). A pH value less than 7 is acidic. A pH value

greater than 7 is alkaline or base. A solution with a pH value of 4 is ten times more acidic than a solution with a pH value of 5, and one hundred times more acidic than solution with a pH value of 6.

The level of acid rain varies greatly around the country (Figure 6–23). For example, South Carolina is reported to be one of the most acidic states in the nation. In Los Angeles, fog has been measured to have the acidic strength of lemon juice.

Rainfall in the northeastern states is extremely corrosive to car paints and finishes. For example, the average pH of rainfall in New Jersey is an acidic 4.3. General Motors now has clauses in some of its new car warranties that exempt them from liabilities involving paint damage in low pH areas.

Acid rain damage generally occurs to the paint pigments, with lead-based pigments the most susceptible. Typically, the damage looks like water droplets that have dried on the paint and caused discoloration. Sometimes the damage appears as a white ring with a clear, dull center. Severe cases show pitting. Discoloration varies depending on the color. For example, acid rain damage to a yellow finish might appear as a white or dark brown spot. Medium blue might have a whitening look. White might be discolored pink, and medium red, purple.

Metallic finishes can be damaged because the acidic solution reacts with the aluminum particles and etches away the finish. A fresh finish is more easily damaged than an aged finish. Lacquers and uncatalyzed enamel finishes are most susceptible to damage, followed closely by catalyzed enamels.

Clear-coated finishes add a layer of protection against acid rain, so vehicles with two- and three-coat finishes are less susceptible to damage. A clear coat protects the paint pigments from discoloration, but it is still possible for acid rain to create a peripheral etch or ring on the clear coat.

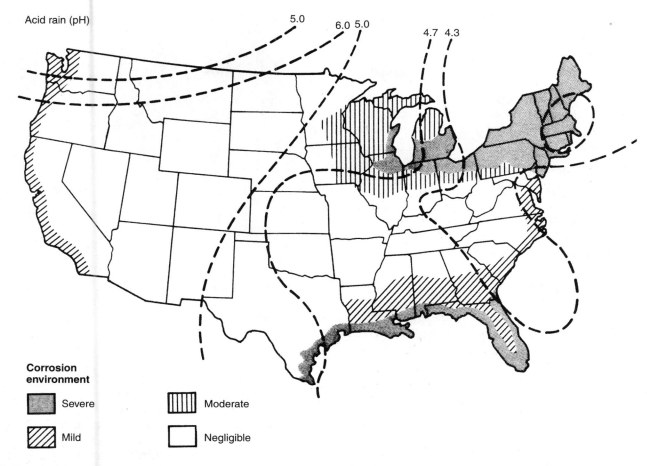

Acid rain (pH)

Corrosion environment

Severe

Moderate

Mild

Negligible

FIGURE 6–23 Notice how acid rain is severe in the northeast and southeast areas of the United States.

RESTORING ACID-RAIN-DAMAGED SURFACES

The procedure for restoring acid-rain-damaged surfaces varies depending on the level and depth of the damage. The following steps outline repair procedures according to the level of damage as illustrated in Figure 6–24. When the problem has been corrected, stop at that stage. Remember that polishing or compounding removes part of the original finish and thereby reduces its overall life. Check paint thickness before buffing. For base coat clear coat colors there should be at least 4 mils of paint. For nonclear coat colors there should be at least 3 mils of paint. If these minimiums are not available refinishing instead of buffing, will be needed.

If the surface damage is like that shown in Figure 6–24A, proceed as follows:

1. Wash with soap and water.
2. Clean with wax and grease remover.

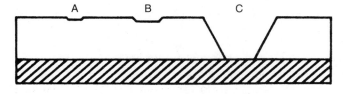

FIGURE 6–24 Levels of acid rain damage.

3. Neutralize the area by washing with baking soda solution (one tablespoon baking soda to one quart of water) and rinse thoroughly.

If the damage is embedded in the surface coat (Figure 6–24B), proceed as follows:

1. Follow cleaning and neutralizing steps already listed.
2. Hand-polish problem area (inspect and continue if necessary).

3. Buff with polishing pad (inspect frequently and remove as little of the original finish as possible to cure the problem).
4. Use rubbing compound (inspect and continue if necessary).
5. Wet sand with 1,500- or 2,000-grit sandpaper and compound. If damage is still visible, repeat with 1,200 grit. Do not use grits coarser than 1,000.

If the damage is through to the undercoat (Figure 6–24C), proceed as follows:

1. Follow cleaning and neutralizing steps listed for Figure 6–24A.
2. Sand with 400- to 600-grit sandpaper.
3. Reclean and reneutralize prior to priming and repainting.

FIGURE 6–25 *During the mounting of a deck lid luggage rack, these holes must be treated to prevent corrosion.*

6.6 SURFACE DAMAGE FROM INDUSTRIAL FALLOUT

Generally speaking, damage from industrial fallout is caused when small, airborne particles of iron fall and stick to the vehicle's surface. The iron can eventually eat through the paint, causing the base metal to rust. Sometimes the damage is easier to feel than see. Sweeping a hand across the apparent damage will likely reveal a gritty or bumpy surface. Rust-colored spots might be visible, however, on light-colored vehicles.

The steps for repairing damage caused by industrial fallout are similar to those used when repairing acid rain damage, but with the following exception: After washing the car, treat the repair area with a *fallout remover,* a chemical treatment product made especially for industrial fallout damage. Do not buff the damaged area before removing the fallout, because buffing will drive the particles into the paint surface. If the particles break loose and become lodged in the buffing pad, deep gouges can occur.

6.7 EXTERIOR ACCESSORIES

To prevent corrosion, it is very important to install a barrier between dissimilar metal components such as aluminum bumpers and stainless and aluminum body trim. The plastic or rubber isolating pads accomplish

this effectively. Mounting stainless and aluminum body trim must be done correctly to avoid galvanic corrosion. Galvanic corrosion happens when two different metals, such as aluminum and steel, are bolted or riveted together. Exterior accessories may be bolted or riveted to the vehicle with steel fasteners. Galvanic corrosion is not a concern if all components are steel. When mounting trim requires drilling holes in the new or repaired panel, drill all holes (Figure 6–25) before applying primer. This way the edges of the holes are protected. If the holes are drilled after painting and not primed, the bare metal edge of the hole will rust.

REVIEW QUESTIONS

1. Technician A says that corrosion requires oxygen. Technician B says that corrosion requires an electrolyte. Who is right?
   a. technician A
   b. technician B
   c. both technician A and technician B
   d. neither technician A nor technician B
2. Galvanized steel has a zinc coating.
   a. True
   b. False
3. Technician A says that welding destroys the corrosion protection of a panel. Technician B

says that welded areas will rapidly rust unless treated. Who is right?
a. technician A
b. technician B
c. both technician A and technician B
d. neither technician A nor technician B

4. Which of the following are conditions that influence the rate of corrosion?
a. air pollution
b. relative humidity
c. temperature
d. all of the above

5. Technician A uses epoxy primer inside a spliced frame rail. Technician B uses undercoat over bare metal. Who is right?
a. technician A
b. technician B
c. both technician A and technician B
d. neither technician A nor technician B

6. Weld scale should be removed with a

_____.

7. Technician A says that acid rain is most severe in the northeastern part of the United States. Technician B says that the western part of the

United States has mild or negligible acid rain. Who is right?
a. technician A
b. technician B
c. both technician A and technician B
d. neither technician A nor technician B

8. A pH value of 3 is an
a. acid.
b. base.
c. alkaline.
d. all of the above

9. Technician A uses thin-bodied seam sealer on one-quarter-inch cracks. Technician B uses brushable seam sealer on smooth finish seams. Who is right?
a. technician A
b. technician B
c. both technician A and technician B
d. neither technician A nor technician B

10. Acid rain damage to paint can be neutralized with baking soda.
a. True
b. False

Chapter

7

Automotive Refinishing Materials

Objectives

After reading this chapter, you should be able to:
- Explain the uses and properties of paints used in the trade for undercoats and topcoats.
- Define the four components of automotive paints.
- Define the characteristics of a good surfacer.
- Identify the types of refinishing.
- Explain the functions of the three types of undercoats.
- Name the types of topcoats.
- Discuss the advances made in refinishing by basecoat/clear coat finishes.
- Explain the advantages of basecoat/clear coat finishes.
- Describe the role of solvents and the variables that affect their spraying.
- Determine the proper solvent to be used for a particular paint job.

Key Term List

orange peel
lacquer
enamel
polychromatic
hardener
reducer
thinner
pot life
flash off
hydrocarbon
organic
crosslink

Automobile finishes perform four very important functions:

- **Protection.** The automobile is constructed primarily of steel sheet metal. If this steel were left uncovered, the reaction of oxygen and moisture in the air would cause it to rust. Painting prevents the occurrence of rust, therefore protecting the body.
- **Appearance improvement.** The shape of the body is made up of several types of surfaces and lines, such as elevated surfaces, flat planes, curved surfaces, straight and curved lines, and so forth. Therefore, another objective of painting is to improve the body's appearance by giving it a three-dimensional color effect.
- **Value upgrading.** When comparing two vehicles of identical shape and performance capabilities, the one with the most beautiful paint finish will have a higher market value. Hence another object of painting is to upgrade the value of the product.
- **Color designation.** Still another objective of painting automobiles is to make them easily distinguishable by the application of certain

colors or markings. Examples are police and fire department vehicles.

The typical automobile finishing system consists of several coats of two or more different materials:

- Undercoat
- Topcoat (color coat or basecoat/clear coat)

The undercoat provides a sound foundation for the topcoat and allows it to adhere better. If topcoats were applied to bare substrates (metal, fiberglass, or plastic), they might peel or look rough; that is why the undercoat is "sandwiched" between the substrate and the topcoat. The undercoat also protects against rust and fills scratches and other flaws in the metal or plastic.

The topcoat is the finish that is seen on the car. From an appearance standpoint, it is smooth, glossy, and eyecatching. Functionally, it is tough and durable. The topcoat thickness on a new car when it comes from the factory is only about 2.5 mils.

A refinisher should know the uses and properties of all paints used in the trade—both undercoats and topcoats. That knowledge will help in choosing the best refinishing system for each job.

Refinishing paints are complex. Those applied to automobiles at the factory have been changing through the years. Manufacturers use high-solids finishes and clear coating. Mica is used in certain topcoats to provide a pearlescent finish. Tricoats—a base color, intermediate pearl coat, and clear coat—are used on some vehicles. Some paints change color depending on the angle at which they are viewed. In more fuel-efficient cars, body parts that were once steel are now manufactured from one or more of the many choices of plastics available to car manufacturers. Each type of plastic exhibits its own characteristics and differences that determine its repairability and refinishing requirements.

In order to provide the perfect matches demanded by the customer on a refinishing job, the painter (Figure 7–1) must respond to these changes. Keeping up-to-date on the changes is crucial. The paint manufacturers who provide finishes for automotive refinishing have quickly responded to the car manufacturers' changes and are getting products to the painter that will enable the painter to effectively repair new cars. In addition to keeping informed and using the right products, the bottom line of a good paint job is the skill and care of the painter in the refinishing process. The techniques employed can vary color, cause loss of adhesion, cause sand scratch swelling, and create a multitude of other problems.

FIGURE 7–1 *Painted car in a spray booth.*

7.1 — TYPES OF PAINT

Vehicles can be painted with one of these three types of paint:

- Lacquer
- Enamel
- Water-based

Paints and solvents may be described as organic compounds. An organic compound is composed of carbon atoms. When carbon atoms are joined with hydrogen atoms, the group of atoms or molecule is known as a hydrocarbon. One characteristic of many hydrocarbons is that they change from a liquid to a vapor at room temperature. Because hydrocarbons evaporate readily, they are described as volatile. Volatile organic compounds (VOCs) are simply carbon-containing molecules that evaporate at room temperature. As discussed in Chapter 1, the release of VOC into the atmosphere causes air pollution. When vehicles are painted, the hydrocarbon solvents used to dilute most paints are released into the atmosphere (by both paint spraying and drying). There are limits on the amount of VOC allowed in the paint. Some heavily populated areas of the United States have more strict regulations on the amount of VOC that may be released during refinishing. These restricted areas must use VOC-compliant paint and HVLP spray guns. Lac-

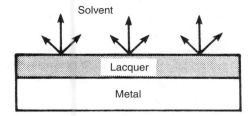

FIGURE 7–2 Lacquers dry by evaporation.

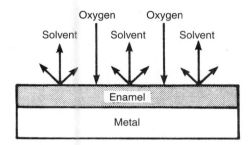

FIGURE 7–3 Enamels dry by evaporation and oxidation.

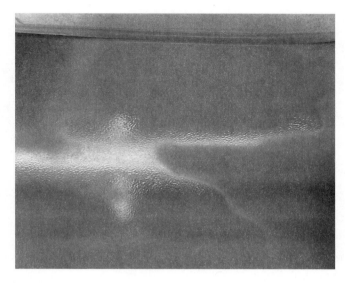

FIGURE 7–4 Orange peel.

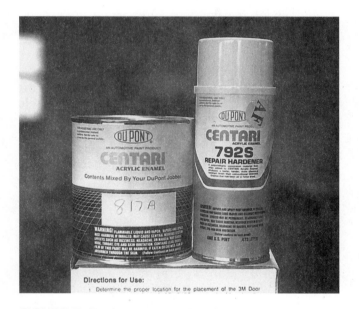

FIGURE 7–5 Acrylic enamel and hardener.

quer paint releases many VOCs. Some enamel paints are VOC compliant, meaning that even though they contain VOCs, the VOC amount is low enough that it does not cause a problem. Manufacturers have modified some enamel paints to make them VOC compliant. Paint company research led to the development of water-based paints for vehicle refinishing. These water-based paints release much fewer VOCs than hydrocarbon solvent-based paints. Water-based undercoats and, from some manufacturers, basecoats are available.

Lacquer paint has been used by both OEMs and refinishers. By the late 1980s, OEMs stopped using lacquer paint. The vast majority of refinishers do not use lacquer at this time. Lacquer paint dries by the evaporation of the solvent (Figure 7–2). The surface dries quickly, but it takes a long time for the paint layer to dry all the way through. The surface must be compounded to achieve gloss. Dried lacquers are not very resistant to solvents, acid rain or ultraviolet (UV) light. Even dried lacquer paint can be turned to liquid if it encounters thinner or other strong solvents. Lacquer paints do not meet VOC regulations and may not be available in the future.

Enamels dry by the evaporation of the solvents as the first stage and by the oxidation of the binder as the second stage (Figure 7–3). This curing results in a change in the binder when it combines with the oxygen in the air. Heat makes these actions more rapid. Thoroughly dry synthetic films are quite insoluble in

ordinary solvents. The longer the drying period, the more insoluble and tougher they become. A thorough sanding prior to repainting is therefore necessary. Enamel finishes dry with a gloss and do not require rubbing or polishing. Since enamels generally dry more slowly than lacquers, there is more of a chance for dirt and dust to stick in the finish. Also, enamel finishes often dry with a texture called **orange peel** (Figure 7–4). While there is generally a slight amount of orange peel in an enamel film, too much will cause surface roughness or lower gloss. Hardener or catalyst is added to the enamel to increase durability and to speed cure time (Figure 7–5).

Water-based paints dry by evaporation of the solvent water. This process can be quite slow, especially in a damp, cool spray booth. A short-wave infrared heater can be used to speed dry time. Some OEMs at certain factories have used water-based paint. Water-based undercoats, such as sealer, can be sprayed on a sensitive substrate to prevents problems with the topcoat. If VOC limits become more restrictive in the future, water-based paints may become more common. At this time, their use is limited.

SHOP TALK

Follow these suggestions when working with water-based paint:

* *Keep the paint from freezing.*
* *Do not use a paint shaker to mix; it will cause foaming.*
* *Flush spray guns with water to remove hydrocarbon solvent before spraying water-based paint.*
* *After spraying water-based paint, clean spray gun with water, then lacquer thinner.*

The majority of today's vehicles are painted at the factory with high-solids basecoat/clear coat enamel finishes. These finishes are baked in huge ovens to shorten the drying times and cure the paint. Very few, if any, refinishing shops have these large bake ovens, but even if they did, the high temperatures might damage a car's upholstery, glass, or wiring. Therefore, refinishing shops require an easy-to-apply, yet durable enamel that will:

* Air-dry without baking
* Dry when baked at low temperatures for thirty minutes at 140 degrees.

More information on the various lacquers and enamels used as topcoats in the automobile industry is given later in this chapter. In addition to the topcoat, there are other materials that play an important role in the refinishing operation.

 7.2 — **CONTENTS OF PAINT**

Automotive paints are composed mainly of three or four components:

* Pigment/metallic or pearl flakes
* Binder
* Solvent
* Additives (with some finishes)

PIGMENT

Pigment is one of two nonvolatile film-forming ingredients (that part which remains in the dried film) found in paint. It provides the color and durability of the finish. It also gives the paint the ability to hide what is underneath. In addition to providing durability and hiding, pigment can also improve the paint's strength and adhesion, change gloss, and modify flow and application properties.

The size and shape of pigment particles are important, too. Pigment particle size affects hiding ability, while pigment shape affects strength. Pigment particles can be nearly spherical or rod- or plate-like. Rod-shaped particles, for example, reinforce paint film like iron bars in concrete.

BINDER

The other nonvolatile film-forming paint ingredient, the binder, holds the pigment in liquid form, makes it durable, and gives it the ability to stick to the surface. The binder is the backbone of paint.

The binder is generally made of a natural resin such as rosin, drying oils like linseed or cottonseed, or a synthetic resin such as methyl methacrylate, polyurethane, urethane, or polyester. The binder dictates the type of paint to be produced because it contains the drying mechanism.

Binder is usually modified with plasticizers and catalysts. These ingredients improve such properties as durability, adhesion, corrosion resistance, marresistance, and flexibility.

SOLVENT

The solvent (or *vehicle* as it is sometimes called) is the volatile ingredient of the paint. Most solvents are derived from crude oil. The main function of a volatile in paint is to make it possible to properly apply the material, and it must be of sufficient solvent power to dissolve the binder portion of the film. High-quality solvents improve the application and film properties of topcoats. They also enhance gloss and minimize paint texture, so less buffing is needed. They also help with more accurate color matches.

In addition to the solvent already in paint, so-called diluent solvents are used to give the paint a viscosity that makes it easier to apply. When used with lacquer, the diluent solvent is called a **thinner**. When used with enamel, it is called a **reducer**. That is an important distinction in the automobile refinishing business, and the respective products are so labeled. Remember that a lacquer is *thinned*, while an enamel is *reduced*. The ratio of pigment/binder/solvent that makes

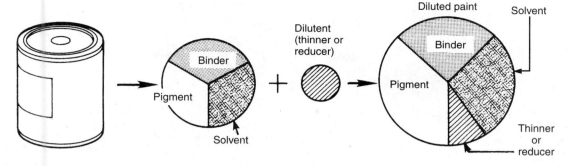

FIGURE 7–6 Ratio of pigment/binder/solvent that makes up paint.

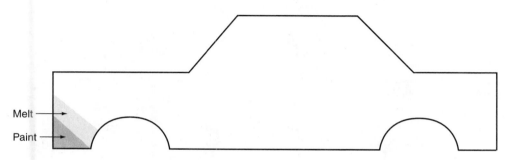

FIGURE 7–7 Spot painting.

up paint is given in Figure 7–6. Of course, the primary solvent in water-based paint is water.

ADDITIVES

With the extensive changes made in paint technology during the last decade, additives have become a way of life for most refinishers. While comprising no more than 5 percent of the paint at most (and usually much less), additives perform a variety of vital functions. Some speed up drying and improve gloss; others slow drying; still others lower gloss. And some perform a combination of functions—such as eliminating wrinkling, providing faster through-cure (the final drying), preventing blushing (a milky, misty look), and improving chemical resistance.

Those additives that speed up cure and improve gloss are often referred to as **hardeners**. Those that slow drying are called retarders. And those that lower gloss are called flatteners.

Flexible paint additives are used in various color coat systems to provide the necessary elasticity to an otherwise rigid paint coating. When a flexible part is compressed or crinkled, the part will return to its normal shape. These additives allow the paint system to flex with the part.

7.3 TYPES OF REFINISHING

With a wide number of refinishing paint materials available, the refinisher uses them to do only one of three types of jobs:

- Spot refinishing repairs (Figure 7–7)
- Panel refinishing repairs (Figure 7–8)
- Overall repainting of the entire vehicle (Figure 7–9)

SPOT REPAIR

This type of repair is often called "ding and dent" work. That is because the damaged area is usually small—either a ding or a dent. Other possibilities might be scratches or a breakdown of the substrate due to rust or corrosion.

Spot repair generally involves:

- Minor body repair.
- Metal treatment.
- Undercoat applications.
- Topcoat application that is blended into the old finish surrounding the repair.

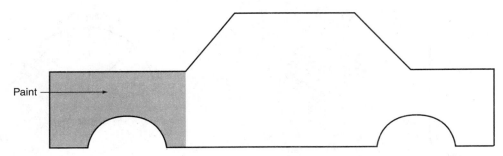

FIGURE 7–8 Panel painting.

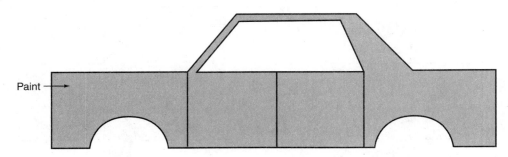

FIGURE 7–9 Overall painting.

For basecoat/clear coat applications, most repairs of this type involve spotting in the basecoat and clear coating the entire panel.

PANEL REPAIR

Panel repair work uses the same technique as spot repair except that the area covers an entire panel or panels of the car (door, hood, and so on) and the color match is made at the panel joints. As in spot repair, any imperfections within the panel must be featheredged and the undercoats applied before the topcoats go on. To prevent color match problems, it is recommended that the new color be blended into the adjoining panels.

Both spot and panel repair could be considered small area finishing. Oftentimes the refinisher must decide whether to do a spot or a panel repair. If the area to be repainted is large, panel repair can be easier.

Note: The majority of paint work done by collision repair facilities is spot and panel repair.

OVERALL REPAINTING

Overall repainting means that the whole vehicle is painted. It is also known as a "complete." Some reasons for refinishing the entire car include:

- Size and/or number of spots to be repaired
- Dull, cracked, or worn finish
- Car owner wishes to change color

There are also specialty paint shops that do mostly overall painting as well as custom shops that develop glamour finishes for custom, antique, and classic cars.

SHOP TALK

Paint manufacturers want the users of their products to be successful. All paint manufacturers publish manuals on how to use their products. Paint company representatives are available to demonstrate new products and to answer questions. All painters should read and understand the paint product manual. If there are problems, the paint company representative should be contracted.

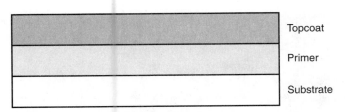

FIGURE 7–10 *Primer over smooth metal.*

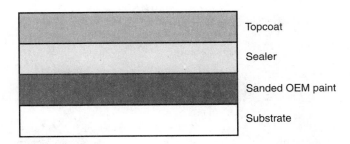

FIGURE 7–12 *Sealer over OEM paint.*

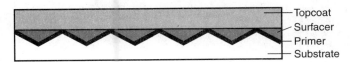

FIGURE 7–11 *Primer and surfacer over rough metal.*

All the types of paint sprayed on to a vehicle are important. The first types of paint applied are the undercoats.

TABLE 7–1: FUNCTIONS OF UNDERCOATS

Name	Purpose	Use
Primer	Corrosion prevention, adhesion	Applied to panel surface
Surfacer	Filler	Intermediate between primer and sealer
Sealer	Uniform surface color and adhesion	Intermediate between surfacer and topcoat
Topcoat	Upgrades the external appearance	Gives color, gloss, and body to help upgrade merchandising value

7.4 UNDERCOAT PRODUCTS

Proper undercoating is part of the foundation for an attractive, durable topcoat. If the undercoat or combination of undercoats is not correct, the topcoat appearance will suffer and might even crack or peel.

Undercoating can be compared to a sandwich filler that holds two slices of bread together. The bottom slice of bread is called the substrate (or surface of the vehicle). That surface can be bare metal or plastic, or a painted or pre-primed surface. The undercoat is the sandwich filler applied to the substrate. It makes the substrate smooth and provides a bond for the topcoat. The upper slice is the topcoat, the final color coat that the customer sees.

Undercoats contain pigment, binder, and solvent. There are three basic types of undercoats:

- Primer (Figure 7–10)
- Surfacer (Figure 7–11)
- Sealer (Figure 7–12)

Most surfaces must be undercoated (Table 7-1) before refinishing for several reasons: to fill scratches to provide a good base for topcoat, to promote adhesion of the topcoat to the substrate, and to assure corrosion resistance. An epoxy primer or self-etching primer alone will not fill sand scratches or other surface flaws. Surfacers are used to fill sand scratches and other minor surface defects. Sealers are applied to promote adhesion and provide a uniform surface for the topcoat. These three undercoats—primer, surfacer, and sealer—can be used together, or in various combinations, depending on the surface condition and size of the job.

With the great number of different undercoat products on the market, the refinisher is often faced with the problem of which to use. No matter which type of undercoat product is used, the golden rule for selecting surface preparation and all other refinish products is the same: Never mix different manufacturers' products.

Refinish products are formulated to work as systems. Manufacturers spend millions of dollars in research and development to design systems of products that work together to provide a specific result. Mixing one manufacturer's reducer with another manufacturer's surfacer or topcoat is almost sure to create a lot of headaches. Putting one manufacturer's sealer over another's surfacer is just as risky. It is important to remember that manufacturers only test and guarantee their own systems.

FIGURE 7–13 *Primer is sprayed over bare metal.*

FIGURE 7–14 *Epoxy primer sprayed inside a frame rail.*

PRIMERS

By definition, a primer is generally the first coat in any finishing system (Figure 7–13). It is designed to prepare the bare substrate and to accept and hold the color topcoat. Primers should provide maximum adhesion to the surface and produce a corrosion-resistant foundation.

There are several special primers available to the shop refinisher. For example, the two-part **epoxy primers** are probably the most versatile and valuable primer products on the market today. An epoxy primer system consists of the primer itself and an activator (or catalyst). When the two parts are mixed together (known as **catalyzing**), they may require an induction time—that is, the period of time that the components must be allowed to thoroughly mix with each other before spraying. It must be remembered that these two-part or two-component mixtures have a **pot life**—that is, a limited time before they become unusable.

Using a two-part epoxy primer is one of the best ways to restore corrosion prevention to a damaged vehicle. The epoxy primers are easy to apply (Figure 7–14) and offer the extra protection that vehicles require because of the high degree of wear and tear to which they are exposed. Virtually any automotive or commercial vehicle topcoat can be used over an epoxy primer. Epoxy primers are available in a variety of colors. Body filler can be applied over cured epoxy primer resulting in the greatest degree of corrosion protection.

Zinc chromate primer has been around for many years. It is a primer designed mainly for adhesion to aluminum. Today, however, epoxy primer will actually do a better job in this application.

Zinc chromate primer should always be used in applications where dissimilar metals will come in contact with each other, such as in a truck body where steel ribs might be used to support aluminum sides and top. One thin coat of zinc chromate will prevent an electrolysis reaction, which causes rapid corrosion. Most zinc chromate primers are reduced with enamel reducer and can be topcoated only with enamels.

Zinc weld-through primer is a zinc-rich material that is frequently used to protect welded joints against corrosion. The zinc in the primer will provide galvanic protection in the weld zone.

When corrosion resistance is very important, **self-etching primers** (also called **vinyl wash-primers**) are used under surfacers or sealers. The reducer in a self-etching primer contains phosphoric acid that "bites" into the surface to provide superior adhesion to steel, aluminum, rigid and flexible plastics, and certain kinds of zinc-coated metal. The bond is so strong, in fact, that when a primer has been applied to bare metal it is extremely difficult to remove without grinding it off.

Self-etching primer must be reduced with a special reducer-catalyst and must be used within a specific period of time. It offers high corrosion resistance to alkalies, oils, grease, and even salt water. Another benefit of this type of resin primer is that it is more forgiving of metal cleaning problems.

Chip-resistant primers are specially formulated for use on lower body sections that are prone to gravel and stone impact damage (Figure 7–15). They give improved resistance against corrosion and help to reduce drumming noise.

Adhesion to plastic panels that have come from the factory unprimed can be improved by the use of a special **plastic primer.** Flexible plastics in particular require plastic primer (Figure 7–16). Many rigid plastics also benefit from priming with a plastic primer. Plastic primers promote bonding and eliminate peeling and other adhesion problems.

FIGURE 7–15 Chip-resistant coating.

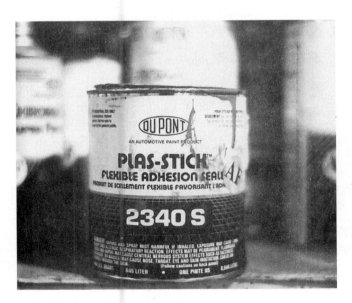

FIGURE 7–16 Plastic primer.

FIGURE 7–17 After the primer is sprayed, a surfacer will be applied to fill in the featheredged area.

SHOP TALK

Most metal replacement parts are supplied by the car manufacturers with a factory primed surface known as E coat. E coat means electrodeposition primer. This E coat has been baked onto the part and is an excellent primer. The car manufacturer usually recommends that this primer not be removed because it is an anticorrosive in addition to its paint adhesion properties. It is necessary to apply a primer or sealer over scuff sanded E coat.

Another ready-to-spray primer product designed specifically for use with basecoat/clear coat OEM finishes is the so-called **adhesion promoter** or **midcoat primer.** This product is a water-clear primer with good durability and excellent adhesion to the very hard clear coats. It is recommended some paint manufacturers require that an adhesion promoter be applied beyond the repaired area on a spot repair. Its purpose is to provide a surface to which a blend edge can adhere.

SURFACERS

Surfacers are the most popular of all the undercoats in refinish applications. They are used to build up featheredged areas for rough surfaces and to provide a smooth base for topcoats (Figure 7–17). A

FIGURE 7-18 Block sanding surfacer with a flat stick and 320-grit sandpaper.

peratures and in a very clean shop environment. An acrylic urethane primer-surfacer is the only choice when a long-lasting, superior paint job is the goal.

All surfacers should be mixed according to label directions and applied as specified, allowing the recommended flash time between coats. Applying heavy coats or not allowing enough flash time will result in a thick coat of surfacer that can gum up the sandpaper, featheredge poorly, and cause loss of gloss or peeling of the topcoat. Excessive thickness of a surfacer coat under a topcoat could lead to premature crazing and/or cracking conditions.

SHOP TALK

A container of hardener should be labeled with the date when it is opened. After a period of seven days, moisture from the air will contaminate the hardener, even in a sealed container. Properly dispose of outdated hardener. Using contaminated hardener in surfacer could lead to hard sanding, sand paper loading, and even lifting of the topcoat.

good surfacer should have all of the following five characteristics:

- **Adhesion.** A strong bond between the primer or factory E-coat and the topcoat to be applied.
- **Build.** A quality that provides the necessary fill for grinder marks and sand scratches in repair work.
- **Sanding ease.** A characteristic that allows the surfacer to be sanded smooth and leveled quickly and easily (Figure 7-18).
- **Holdout.** A sealing quality that prevents the topcoat from sinking into the surfacer, resulting in a dull look.
- **Drying speed.** A time-saving quality that permits the refinisher to go on to the next operation.

Though fairly new on the market, **self-etching surfacers** are rapidly gaining in popularity because in one step they etch the bare metal and provide the fill needed for a smooth finish. Etching filler ensures outstanding paint adhesion and corrosion resistance and can reduce surface preparation time and materials costs by more than 25 percent.

The three-component, surfacer, hardener and reducer, **acrylic urethane surfacers** were created for use with today's more sophisticated and expensive topcoats. Acrylic urethane surfacers provide high film build, color holdout, and minimal film shrinkage, substantially reducing the risk of overnight dull-back. The tradeoffs are higher cost, overspray damage, and the need to apply these products at controlled shop tem-

SEALERS

Sealers are sprayed over a primer, a surfacer, or a sanded existing finish. They are used in automotive refinishing for four specific purposes:

- They offer better adhesion between the paint material to be applied and the repair surface (Figure 7-19). Sanding of the surface is usually required before application.
- They may act as a barrier-type material that prevents or retards the mass penetration of refinish solvents into the color and/or undercoat being repaired.
- They provide uniform holdout. If the old finish is good and hard and if a surfacer with good holdout is used for spot repairing, a sealer is not mandatory. Obviously, if only one or neither of these conditions is present, a sealer is recommended. However, a sealer is always recommended to provide uniform color holdout and to prevent die-back.
- They provide uniform color. Uniform color is useful when spraying a poor hiding color, such as red pearl. A dark red sealer will prevent showthrough of a light-colored surfacer in this case.

Sealers are available in different colors to allow hiding with a minimum number of topcoats. Typical colors are black, white, dark red, gray, and beige. Other types of sealers may be tinted up to 5 percent

FIGURE 7–19 *A sealer is sprayed on a replacement fender.*

with topcoat, so the sealer is the same shade as the topcoat. A paint system is available that uses combinations of black and white sealer to produce various shades of gray. The topcoat formula specifies which sealer combination will give the best hiding.

There are situations in which some kind of sealer must be used. Check paint manufacturers' recommendations on when sealer should be used. The following are some examples of when a sealer should be used:

- over scuffed E coat on a replacement part—for adhesion of topcoat
- over sanded OEM paint on a complete refinish— for adhesion of topcoat
- over a single panel or series of panels that have areas of contrasting color surfacer—to provide a uniform color

SHOP TALK

The phrase "wet on wet sealer" means that the sealer is sprayed, and after proper flash time, the topcoat is sprayed over the sealer. The sealer is not sanded before topcoat. However, if dirt specks are present in the sealer, they may be removed by light sanding.

In summary, the selection of one or more undercoats will be determined by the following characteristics of a job:

- **Type of surface.** Bare metal or previously finished.

- **Condition of that surface.** Repaired area or sanded aged finish.
- Whether **holdout** and **show-through** are problems.

Depending upon these circumstances the painter must choose one of the following:

- Primer by itself
- Surfacer by itself
- Sealer by itself
- Primer with a surfacer
- Surfacer with a sealer
- Primer, surfacer, and sealer

No matter what type of refinish job confronts the painter, the purpose of the undercoat is always the same: to provide the best possible foundation for a beautiful, long-lasting finish. Making the right product choices is an important step in paving the way for a successful paint finish.

7.5 — TOPCOATS

From the customer's standpoint, the topcoat (or color coat) is the most important in refinishing because that is all the customer sees. The expert refinisher takes special pride in producing a beautiful finish on spot and panel repairs, or in an overall repaint job that matches both the color (or color effect) and the

FIGURE 7–20   In metallic finishes, the light enters the finish and is reflected by metal flakes to produce a metallic color effect. (Courtesy of Du Pont Co.)

FIGURE 7–21   Dry spray traps the metallic particles at various angles near the surface and causes a high-metallic color effect. (Courtesy of Du Pont Co.)

FIGURE 7–22   Wet spray allows time for metallic particles to settle in the paint film and causes a strong pigment color effect. (Courtesy of Du Pont Co.)

texture of the original finish. Therefore, it is of great importance to fully understand the topcoat materials and how they are applied.

AUTOMOBILE COLORS

Like all colors, the appearance of automotive refinishing colors is a result of the way they react to light. The color seen is the result of the kind and amount of light waves the surface reflects. When these light waves strike the retina, they are converted into electrical impulses that the brain sees as color. The same color will look very different under natural daylight, incandescent lamps, or fluorescent bulbs. That is why it is so important to check color match in daylight or a balanced artificial light. There are two basic automobile finish types: solids and metallic colors.

Solid Colors

For many years all cars were solid colors, such as black, white, tan, blue-green, and maroon. These colors are composed of a high volume of opaque pigments. Opaque pigments block the rays of sun and absorb light in accordance with the type of color they are. That is, the darker the solid color is, the more light it absorbs and the less it reflects. Black will absorb more light and will reflect less; white absorbs less light but reflects a great deal more. When polished, solid colors reflect light in only one direction. Solid colors are still used by the refinisher, but to a lesser degree when compared with a few years ago.

Metallic Colors

Metallic (or polychrome) paint contains small flakes of aluminum suspended in liquid. Pearl paint has mica particles in it. The metal particles combine with the pigment to impart varying color effects. The effect depends on the position the flakes assume within the paint film (Figure 7–20). The position of the metal flakes and the thickness of the paint affect the overall color of the painted surface. The flakes reflect light, but some light is absorbed by the paint. The thicker the layer of paint, the greater the light absorption.

When metallic paint is sprayed on dry, the metallic flakes are trapped at various angles near the surface. Light reflection is not uniform, and because the light has less film to travel through, little of it is absorbed (Figure 7–21). The result of nonuniform light reflection and a minimum of light absorption is a painted surface with a metallic appearance and a light color.

When metallic paint is sprayed on wet, the metallic flakes have sufficient time to settle, so they lie parallel to and deeper within the paint film. Light reflection is uniform and, because the light has to go farther into the paint film, light absorption is greater (Figure 7–22). The result is a painted surface that appears deeper and dark in color.

CAUTION: Metallic color paints must be stirred and mixed thoroughly before use. As shown in Figure 7–23, the pigment settles below the binder; the metal flakes settle below the pigment. If flakes stay at the bottom of the can, the paint will not match the same color on a vehicle being repaired or refinished.

BASIC TYPES OF TOPCOATS

Automotive topcoat finishes range from paints that have been available for fifty years to new multicomponent systems that provide the ultimate in durability. No single finish is best for all applications; it all depends on a careful matching of finish capabil-

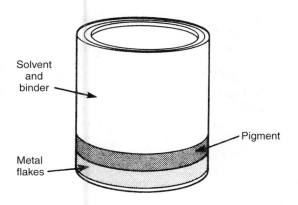

FIGURE 7–23 _Metallic paint is composed of solvent, binder, pigment, and metal flakes. The pigment settles below the solvent and binder, while the flakes sink below the pigment._

ities and characteristics with the requirements of the application.

Here is a brief review of the history and highlights of various materials used as topcoats. The application of those still being employed by the automotive trade is detailed in Chapter 11.

From 1900 to the early 1920s, the wooden bodies and spokes of cars were painted by hand. A mixture of lampblack pigment and varnishes was brushed on by hand, coat after coat, until coverage and hiding were achieved. This pigment, having very few solids and being weak, could not give enough coverage with even five coats. This buildup of material, lacking any modern driers, would take from forty to fifty days to dry.

When the bodies were finally dry enough to work on again, a pumice and damp rag, along with plenty of elbow grease, were applied and any flaws—such as remaining brush marks, dirt, or runs—were removed. No color selection had been developed at this time, nor did this material have any durability. It dulled and cracked within a year after all that hard work.

Nitrocellulose Lacquers

The introduction of nitrocellulose lacquer in 1924 revolutionized automobile finishing. The use of this lacquer, applied with a spray gun, allowed the steel vehicle bodies to be completed in one day, thereby eliminating a major bottleneck in automobile production.

Nitrocellulose lacquer required several coats of color that air-dried very quickly but faded and chalked because of weak pigment. The color range was just

beginning to show improvement, with something other than black, as the greens came into focus. Reds, blues, and yellows were also developing, along with solvents and other technology. Nitrocellulose lacquers were used by the American automobile refinishing trade well into the 1950s.

Alkyd Enamels

Introduced in 1929, alkyd resin enamel has been satisfying refinisher needs in the repainting of passenger cars and commercial vehicles for more than half a century. This finish is basically the same today as it was when it was first introduced, with a slight modification. The name _synthetic_ was given to this enamel finish because the resin used in its formulation was modified by synthesizing the enamels with simple petroleum chemical compounds. Now, of course, all paint resins are synthesized, but the name synthetic enamel is still used to identify these alkyds.

Synthetic or alkyd enamel dries to a lustrous high gloss, requiring no compounding or buffing. It is the least expensive and the least durable of the commercial finishing systems and tends to dry slowly, generally overnight. It is prone to wrinkling in hot weather during the first twenty-four to forty-eight hours, and it cannot be two-toned or repaired for about eight to twenty-four hours. Synthetic enamel also has poorer resistance to water, chemicals, and gasoline during the first few days after application.

Alkyd synthetic enamel is rarely used today. However, a few refinishers still select this enamel because it provides an economical two-coat system with high initial gloss that hides surface imperfections better than newer enamels. This is especially true when a synthetic enamel is reduced with hardener. The synthetic is applied the same as without hardeners, but the drying time is shortened when the hardener has been added. The finish can be two-toned when it is dry enough to tape and generally can be recoated at any time without fear of lifting. The hardener also makes the synthetic enamel highly wrinkle-resistant, though it does not improve durability.

Acrylic Lacquers

Introduced in 1956, acrylic lacquer quickly replaced nitrocellulose lacquer because it dried to a higher gloss and retained its appearance longer. It found immediate acceptance as a favored OEM and repair finish. It is used for spot and panel repair of both lacquer and enamel original finishes and in overall repainting. When used over enamel, a sealer is recommended.

Properly applied, an acrylic lacquer system can give excellent results. It combines the features of

FIGURE 7–24 *Always read the paint label directions before use.*

durability, color and gloss retention, high gloss, and fast drying time. To achieve these features, both the undercoats and the thinners must be those recommended for acrylic lacquer paints. The manufacturer's instructions should always be followed. If products from different manufacturers are mixed, the paint film might crack or craze.

Acrylic lacquers respond readily to hand or machine polishing with rubbing or polishing compounds. Some degree of polishing is usually necessary, depending on the gloss level required.

Because of a lack of durability and noncompliance with VOC limits, lacquers are rarely used in collision repair facilities today.

Acrylic Enamels

Developed in the 1960s, acrylic enamels are more durable and faster drying than alkyd enamels, and they eliminate the compounding step associated with acrylic lacquer. Generally recommended for panel repairs and overall repainting, some acrylic enamels are used for spot repair of both lacquer and enamel original finishes. Acrylic enamel costs slightly more than synthetic enamels but is less expensive than other finishing systems available today.

Acrylic enamel combines the properties and advantages of the very durable color and gloss-retentive acrylic lacquer with the excellent flow characteristics and gloss of the enamel systems. It is fast drying and is about 50 percent more durable than synthetic enamel. It is also easy to apply, produces a deep luster and high gloss, and has excellent color and gloss retention. Two-toning or recoating is possible within the first six hours of drying. During this period, the sol-

vent evaporates from the film, leaving the finish non-tacky and dry to the touch.

For the next six to forty-eight hours, however, most acrylic enamels enter the hard-curing period. Recoating must be done before this period; if it is not, the strong solvents in the new coat can cause lifting of the original coat, because it has become too cured to dissolve and melt in with the new coat. If recoating is necessary during this time, most paint manufacturers recommend the use of a recoat sealer that serves as a carrier coat and permits the recoating of acrylic enamel during the critical cure period when lifting might otherwise occur.

Some of the problems created when using acrylic enamels can be reduced or eliminated by the use of a hardener or urethane catalyst. The addition of a hardener or catalyst (often referred to as a second stage or second component) provides a greater initial gloss than synthetic enamel or acrylic enamel alone. It improves the DOI (distinctness of image), which is a measurement of the sharpness of images reflected from the surface. It also improves the chemical resistance, flexibility, and chip resistance of the finish while eliminating the self-lifting that might occur during recoating within the hard-curing period of acrylic enamel without hardener. When adding a hardener or catalyst to acrylic enamel, a recoat or barrier coat sealer is not necessary to prevent lifting.

After reducing the acrylic enamel color and just prior to spraying, add the catalyst or hardener to the reduced material according to label directions (Figure 7–24). Once the catalyst is added and mixed, the topcoat should be sprayed immediately. The **pot life,** or working time, of these paints will generally be in the range of four to eight hours. After this time, all mixed paints should be discarded.

The advent of ultraviolet absorbers has greatly increased the durability and exterior weathering of acrylic enamel. Acrylic enamel with hardeners is sprayed in the same manner as ordinary acrylic enamel but provides 50 percent greater durability than acrylic enamels and 100 percent greater durability than the alkyd or synthetic enamels.

Polyurethane Enamels

Ordinary polyurethane enamels provide significantly greater durability than alkyd or synthetic enamels, acrylic enamels, and acrylic enamels with hardeners.

The polyurethane enamel—made with a polyester resin—dries by evaporation of the solvents and by chemical cross-linking between the two principal base components (hydroxyls and isocyanates) to cure the paint film. The cross linking results in a high-gloss, extremely durable, chemical- and solvent-resistant finish. Without an accelerator, dry time to tape is about sixteen hours. With an accelerator, dry time to tape can be reduced to approximately three to five hours. Polyurethane enamels are used on heavy duty and semi trucks.

Acrylic Urethane Enamels

Urethane is formulated with an acrylic resin that is much more durable than any other finish. Because of the characteristics of this resin, an acrylic urethane enamel offers higher gloss retention and, ultimately, a much longer life. Recent manufacturer tests have shown that this finish retains its gloss from 50 percent to 100 percent longer than ordinary polyurethane enamels.

The acrylic urethane enamel also dries tack-free and out of dust faster than polyurethane enamels. It is also more resistant to film degradation caused by the sun's ultraviolet (UV) rays. While degradation of the finish cannot be totally stopped, the chemical composition of this material substantially retards damaging UV effects. When compared to metallic polyurethane colors, acrylic urethane enamel metallic colors have twice the durability. The urethane's durability also offers exceptional resistance to stone chipping and does not break down when exposed to weather, gasoline, chemicals, dirt, and road grime. It virtually eliminates sand scratch swelling.

To achieve proper results from this two-component high-quality material, it is necessary to closely follow the manufacturer's instructions to the letter. The pot life of the activated material is one and one-half to two hours. To achieve the best results, spray the material as soon as possible after the components are mixed together. Urethane enamel dries by crosslinking in the same way as polyurethane. Urethanes can be solid or polychromatic—metallic or pearl, single stage, basecoat/clearcoat, or tricoat.

Urethane clear coats can be applied over polyurethane enamel and properly prepared acrylic enamels thirty minutes after the color coat application. The result is a wet look that gives the basecoat additional depth, maximum color and gloss retention, and extended durability.

WARNING: Polyurethane and acrylic urethane enamels as well as the two-component or catalyzed-type finishes (acrylic and alkyd) use isocyanates as hardeners. Isocyanates are highly reactive and can become health hazards if not handled carefully. Because isocyanates react with many common substances, they must be used, stored, and disposed of with care. All urethane paints and products contain isocyanates and, therefore, share the same potential health hazards regardless of brand.

TABLE 7–2: COMPARATIVE DURABILITY OF TOPCOATS

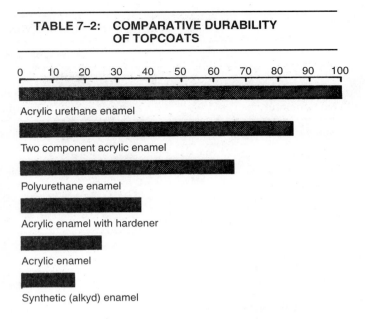

Appropriate measures must be taken to prevent overexposure or sensitization whenever urethane or any other paints containing isocyanates are being sprayed. These paints should only be used in areas with adequate ventilation. For maximum protection when isocyanates are being sprayed, a supplied air respirator should be worn until the work area has been exhausted of all vapors and spray mist.

Tables 7-2 and 7-3 show the relative durability of the various types of topcoat systems and a comparison to original equipment (OE) high-baked thermosetting acrylic enamels.

Costs can be deceiving, particularly with the polyurethane and acrylic urethane colors, both of which are substantially more expensive than the others in their unreduced prices per gallon. When costs are viewed in terms of life cycle—for example, the length of time the finish will continue to present an acceptable appearance before requiring refinishing—acrylic urethane enamels prove to be the most economical.

The type of paint used for the topcoat ultimately determines the attractiveness of the color, gloss, and finish.

Basecoat/Clear Coat Finishes

In the past few years, OEMs around the world have increasingly adopted basecoat/clear coat systems as the finish of choice for new cars rolling off assembly lines. Over 90 percent of current year vehicles have BC/CC. The technology for basecoat/clear coat finishes was developed in Europe. The durability and popularity of these finishes prompted Japanese and

TABLE 7–3:   SUMMARY OF TOPCOAT PAINT FEATURES

Nomenclature		One-Component Type			Two-Component Type	
		Alkyd Enamel	Acrylic Lacquer	Acrylic Enamel	Polyurethane	Acrylic Urethane Enamel
Spray characteristics		Excellent	Excellent	Good	Good	Good
Possible thickness per application		Fair	Fair	Good	Excellent	Excellent
Gloss	without polishing	Fair	Good	Good	Excellent	Excellent
	after polishing	Good	Good	Good	—	Good
Hardness		Good	Good	Good	Excellent	Excellent
Weather resistance (frosting, yellowing)		Fair	Fair	Good	Excellent	Excellent
Gasoline resistance		Fair	Fair	Fair	Excellent	Good
Adhesion		Good	Good	Fair	Excellent	Excellent
Pollutant resistance		Fair	Fair	Fair	Excellent	Excellent
Drying time	to touch	68°F 5-10 minutes	68°F 10 minutes	68°F 10 minutes	68°F 20-30 minutes	68°F 10-20 minutes
	for surface repair	68°F 6 hours 140°F 40 minutes	68°F 8 hours 158°F 30 minutes	68°F 8 hours 158°F 30 minutes	—	68°F 4 hours 158°F 15 minutes
	to let stand outside	68°F 24 hours 140°F 40 minutes	68°F 24 hours 158°F 40 minutes	68°F 24 hours 158°F 40 minutes	68°F 48 hours 158°F 1 hour	68°F 16 hours 158°F 30 minutes

American automobile manufacturers to begin offering them, too. When the system was introduced, it used either an acrylic lacquer clear or a polyurethane enamel clear over an acrylic lacquer basecoat. Early in 1985, the first acrylic enamel basecoat/clear coat refinish system to actually simulate the OEM basecoat/clear coat finish was introduced. Like OEM finishes, the new system loads more transparent pigments into the base coat and locks the metallic flakes into a flat arrangement, enabling it to match the brightness, the color intensity, and the travel of an OEM basecoat/clear coat finish. This makes it easy to get the best color match available.

An added benefit of this new basecoat technology is a thirty-minute recoat time with the clear coat. Conventional basecoats use "chemical drying" to build solvent resistance. Chemical drying generally requires four to six hours to prevent the clear coat solvents from redissolving the basecoat and allowing the aluminum or mica flakes to streak and mottle. Another new basecoat technology builds the necessary solvent resistance without chemical drying in just thirty minutes at 75°F and 1.5 mils of dry film thickness. These qualities enable this new basecoat/clear coat system to be applied quickly and easily with superior results.

Painters have switched to the new acrylic urethane basecoat/clear coat system. All refinish paint manufacturers now offer this excellent two-stage system, and they have quickly achieved a strong following. Its popularity stems from meeting the practical needs of the painter and from its advanced technology.

Acrylic urethanes are fast, they offer the best color matches, and they provide better coverage and hiding. Thus, they increase productivity and improve customer satisfaction.

Most cars probably are finished in basecoat/clear coat colors. This trend has had several effects on the paint shop. Surface preparation and a clean shop environment are more important than ever, as the higher gloss and DOI (distinctness of image) of the clear coat make imperfections easier to see. A greater emphasis on cleanliness will result in more use of downdraft booths in progressive paint shops.

With good surface preparation and a clean shop, however, painters find that the new acrylic urethane enamel basecoat/clear coats are very easy to apply and result in better looking, more durable finishes. For example, painters no longer have to balance flow for metal control in a metallic color, and streaking and mottling are eliminated. Spot repair is made easy because basecoat/clear coat finishes make it simple to blend in an edge. But most important to paint shop refinishers is the outstanding color match that state-of-the-art basecoat/clear coat systems provide.

One challenge the paint shop refinisher faces is becoming well educated about the many different systems available in order to select the best one. No customer likes a streaked finish or color drift. No paint shop can afford to lose potential income because shop time is tied up by difficult application procedures and long dry times. When one is choosing a refinishing system, the key factors involved are appearance and ease of application.

It is a good idea to look for a system that provides fast dry time and locks in the metallic flakes for consistent color match. Also, the amount of pigment in the basecoat should be considered. A good basecoat will contain enough pigment to achieve hiding in 1 to 1-1/2 mils (two coats of basecoat). Some basecoats will require four to six coats, which means more application time, more materials, and longer dry times due to higher film thickness. To assure long-term durability and customer satisfaction, the system should contain light stabilizers and ultraviolet light absorbers.

Identifying cars that are clear coated is easy. Looking at the vehicle identification code and the color chip book is a quick way to find out if the car has the basecoat/clear coat system. If the code has been removed or destroyed, sanding a small spot in a concealed area of the vehicle to be finished, using a fine sandpaper, can help determine the type of finish. If the dust is white, the car has a basecoat/clear coat finish. If the dust is the color of the car, it does not.

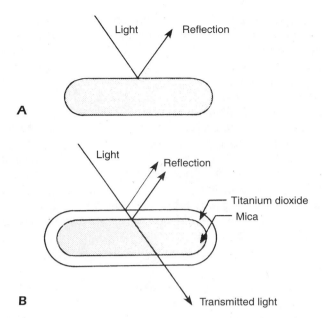

FIGURE 7–25 As opposed to (A) an aluminum flake, which only reflects light off its surface, (B) colored mica particles can be designed to reflect, absorb, and refract differing amounts of light striking them, thereby changing the color.

Pearl Luster Finishes

In their effort to attract buyers in a market where cars are starting to look alike, car manufacturers offer highly iridescent colors applied in two layers. The first stage is a mica or "pearl" coat, and the final stage is the clear topcoat.

In 1960, a synthetic pearl luster pigment using a mica particle covered with thin layers of titanium dioxide was developed. The mica particles made very good carriers for the titanium dioxide because of their highly transparent qualities. The titanium dioxide layers provided the rainbow or pearl effect as light reflected and passed through the layers.

This transparent quality of the pearl luster pigments creates a reflective brilliance that cannot be obtained with aluminum flakes. Aluminum flakes act like miniature mirrors that reflect light (Figure 7–25A). They will not let light pass through to the color. If too many aluminum flakes are added to a brilliant color, the color will be washed out.

Pearl luster pigments of titanium dioxide–covered mica flakes reflect light while also allowing some light to pass through to other mica flakes and colored pigments below (Figure 7–25B). The brilliance and highly iridescent effect of the finish are created this way. Because the pearl coat stage of this system is translucent and reflective, the amount of mica in this stage is critical for matching.

The repaint formulas available for the pearl colors on new cars usually have colored pigments and mica pigments combined. These formulas are applied like any other two-stage paint, except when they are to be sprayed over a repaired area that has been primed. Several additional coats with the proper flash times between them will be necessary because of the transparent quality of the paint. Allow more spraying time when scheduling jobs with mica finishes.

Several things should be kept in mind when working with the pearl luster paints:

- Mica flakes are heavy. Keep the paint agitated to ensure even distribution.
- Spray test panels (with primer spot). Do not test on the customer's car.
- Continually blend.
- Do not rush. Allow enough flash time between coats.
- Spray in a well-lighted booth.
- Ultraviolet lights can help in checking the pearlescent effect.
- Direct sunlight is the best source of light for evaluating touch-ups.
- Check work from three angles in direct sunlight: straight in, from a forty-five-degree angle to the surface with the light behind the observer, and from the opposite forty-five-degree angle with the light ahead of the observer.

Some experimenting with tinting of the base colors, tinting of the pearl coat with colored pigments or pearl pigments, or a combination of both might be necessary to accomplish a good match.

Tricoat Finishes

Tricoating, a three-stage basecoat/clear coat technique, which for thirty years has been used in glamour-coating custom cars and for other special applications, has now found its way into the production line of several manufacturers' deluxe models.

As the name implies, these tricoat finishes consist of three distinct layers that produce a pearlescent appearance: a basecoat, a midcoat, and a clear coat (Figure 7–26). A solid color, such as white or red, is used as the basecoat. The midcoat is a nearly transparent coat containing pearl. The clear coat protects the first two coats. The basecoat is visible through the pearl coat. A tricoat pearl finish has more depth than

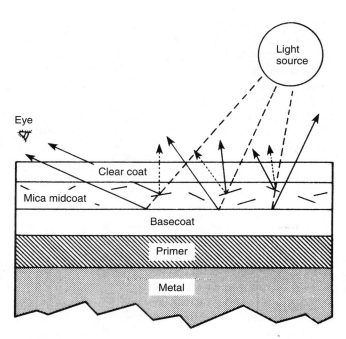

FIGURE 7–26 *A tricoat-solid basecoat, pearl or mica midcoat and clear coat.*

a basecoat/clear coat pearl because there are three distinct layers instead of two. The overall effect of a tricoat is a glowing iridescence.

Hue-Changing Colors

A hue-changing color looks like different colors when viewed from different angles. For example, a hue-changing color may look like dark blue when viewed straight on. From a forty-five degree angle, it may look purple. When viewed from the opposite forty-five degree angle, it may look green. This change in hue is due to special particles in the paint. These particles are not aluminum or mica. The particles are made of polyester and have imprinted grooves. The grooves diffract light, similar to the way a prism diffracts light. Light diffraction separates light into the color spectrum. The appearance of colors from diffraction change as the viewing angle changes. Therefore, the color or hue of a vehicle changes depending on viewing angle. Some vehicles have hue-changing colors applied at the factory. Hue-changing colors are also available for refinishing. The colors may be basecoat/clear coat or tricoat depending on the paint manufacturer. Applying these colors is usually no different from spraying a pearl basecoat/clear coat or tricoat.

FORCE-DRYING ENAMEL TOPCOATS

Force-drying enamels by means of heat convection ovens or infrared lights will greatly reduce the drying period, but care must be exercised to avoid wrinkling, blistering, pinholing, or discoloration. It is generally better to force-dry at lower temperatures for longer periods than to run high temperatures for shorter periods. Consult the paint manufacturers' recommendations for force-drying times and temperatures. Usually a waiting period is required after spraying is completed and before force-drying or baking can begin. This flash time allows solvent to escape from the fresh paint. One type of urethane enamel is flashed for 10 minutes then baked at 140°F for 30 minutes. After the surface cools down, the vehicle can be handled. Baking will greatly increase the productivity of a paint shop.

7.6 — SOLVENTS

The basic function of diluent solvents such as reducers is to lower the viscosity of undercoats and topcoats that are too thick to spray as they come from the container. To accomplish this, a reducer must be properly balanced. That is, it must be made of the right ingredients to:

- provide proper dilution of the material so that it will not only pass through the gun smoothly but also atomize easily as it leaves the gun.
- keep the material in solution long enough for it to flow out and create a smooth, even surface without sagging or running.
- provide complete evaporation of the solvent in order to leave a tough, smooth, and durable film.
- lock metallic flakes in place to prevent clumping or mottling.
- aid in melting clear coat repairs.

WARNING: Solvents such as thinner are also used to clean painting equipment, not the painter. Do not use thinner to remove paint or undercoats from the skin. There are safe products available to remove paint from the skin. Use safe products.

SELECTING THE PROPER SOLVENT

The evaporation of solvent is called flash off. Proper flash-off time is critical. The paint needs to flow out or level after it is sprayed. If the flash-off time is too fast, the paint cannot properly flow. If the flash-off time is too slow, the paint has too much time to flow. Improper flash-off time, either too fast or too slow, may result in problems. There are several interrelated factors that affect flash-off time. Some of these factors will be mentioned briefly here and explained in more detail later in the book. The factors that affect flash-off time are:

- air temperature and humidity
- viscosity of the paint
- thickness of the paint coats
- air flow
- speed of solvent

Here is how temperature and humidity affect flash-off time:

- Hot, dry weather produces the fastest dry time.
- Hot and humid, or warm, dry weather produces a fast dry time, but slower than hot, dry weather. (High humidity can cause problems.)
- Normal weather—70°F with 45 to 55 percent relative humidity—produces a normal dry time.
- Cold, dry weather produces a slower dry time than normal.
- Cold, wet, or humid weather produces the slowest dry time.

The viscosity of the paint is determined by the mixing ratio of paint and reducer. It stands to reason that if excessive reducer is added to paint, the flash-off time will increase. It will take longer for the solvent to evaporate if there is more solvent. All manufacturers list proper reduction ratios for paint. Always read the recommendations and carefully measure the proper amounts of reducer and paint.

The thickness of paint coats or film build affects the flash-off time. Thicker coats require more time to dry. In this case, thick paint could result from spraying one coat and immediately spraying another coat over it. Problems arise if the surface dries before all the solvents in the film evaporate. Follow manufacturers instructions about how much waiting time or flash time to allow for between coats.

Air flow when spraying is also involved in determining flash-off time. More air flow makes the solvent evaporate faster. Paint sprayed on a vehicle in a spray booth with minimal air flow, perhaps due to clogged intake filters, will evaporate slowly. As the paint dries, the solvents evaporate into the air. If the air in the booth becomes contaminated with evaporated solvent, due to poor airflow, the paint will not set up properly. In extreme cases, a lack of free-flowing oxygen results in dull paint.

The last factor in determining flash-off time is the speed of the solvent. Reducers are formulated to evaporate at different rates. The evaporation rate of a reducer is known as speed. Reducers can be described as fast, medium, or slow dry. The speed of the reducer should be matched to the temperature of the spray booth. Properly matched reducer will give the right flash-off time. The following are the different speeds of reducers and the temperatures they should be used at:

- fast dry   60°–70°
- medium dry   70°–80°
- slow dry   80°–90°

This is what one manufacturer recommends. Other paint manufacturers may be different.

If a fast dry reducer is used in a warm booth, the paint will not have enough time to flow out properly. Some problems from the lack of flow include low gloss, orange peel, overspray, and mismatch. The painter may notice the problem that the paint is drying too fast as he sprays and attempt to spray the paint wetter to compensate. This will not work. Spraying the paint wetter by moving the spray gun slower will lead to additional problems such as excessive film build. Cracking may occur as a result of excessive film build.

If a slow dry reducer is used in a cool-temperature booth, the paint will have excessive flow-out time. Excessive flow-out time can lead to paint runs, metallic clumping or mottling, mismatch, and sand scratch swelling. If the painter attempts to spray the paint dry to compensate by moving the spray gun faster, problems such as dulling, mismatch, and lack or durability may result.

Most factors that affect flash-off time can be controlled. Always read and follow the manufacturers' instructions for reducer speed, air flow, film thickness, and paint viscosity. If one or more of these factors is not within specifications, the paint job will suffer.

SHOP TALK

Air flow and temperature must be maintained as the paint dries. Air flow is needed to remove evaporated solvents and provide oxygen. Temperatures above 60°F are required for crosslinking. Loss of gloss or lack of durability will result if airflow is diminished or if the temperature falls below 60°F in the drying area.

7.7 — CLEAR COAT

Urethane clear coat is usually mixed with a hardener and in some paint systems, a reducer. In general, there are three types of clear:

- **High-production clear coat.** This type of clear coat has a two-minute dust-free time. Dust-free time is the amount of time in minutes from when a paint is sprayed until it has dried enough so that dust will not stick to it. Any dust that lands on the clear coat after the dust-free time has passed can be wiped off. Two-minute dust-free time is very fast for an enamel. This high-production-type clear coat also dries rapidly, in 4 hours. This type of clear coat can be difficult to spray because it sets up so fast. High-production clear coat should be used on panels only, not a complete.
- **General use clear coat.** The dust-free time for general use or medium clear coat is ten minutes. Ten minutes is plenty of time for dust to find its way into the wet clear coat. A clean spray booth is needed to make this clear coat look good. Dry time is eight hours or overnight. This clear coat may be used on panels and complete repaints.
- **High-glamour clear coat.** This high-solids clear coat has a long dust-free time, 15–25 minutes. A clean down-draft booth is needed to spray this clear coat. Air dry time is overnight, but baking is the preferred method of drying. This clear coat gives the maximum amount of gloss. It can be used on panels or completes.

The clear coat hardener is the catalyst that causes cross-linking or drying of the clear. When painting a single panel, the drying time can be rapid. When clear

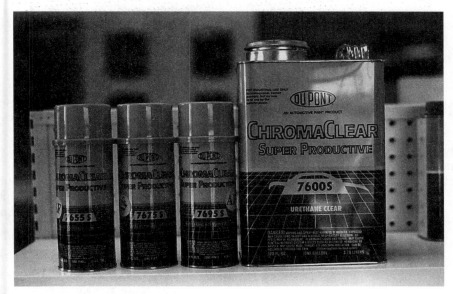

FIGURE 7–27 *Clear coat and three hardeners.*

coating adjacent panels, the clear must dry slower to allow overspray to melt in. If the entire vehicle is clear coated, the dry time must be even slower. The hardener must be matched to the type of repair (Figure 7–27). The hardener may also be matched to spray booth temperature in the same way as reducer is matched to spray booth temperature.

SHOP TALK

As with surfacer hardener, when clear coat hardener is opened, write the date on the can. Once opened, the hardener must be used within seven days. Properly dispose of out of date hardener. Using out of date clear coat hardener may result in loss of gloss, poor adhesion, or improper cure.

7.8 OTHER REFINISHING MATERIALS

Vinyl Spray Colors. Vinyls dry by the evaporation of the volatiles. They dry rapidly to a permanent flexible film. The use of vinyl liquid over vinyl that is cleaned and pretreated with vinyl conditioner results in chemical bonding of the two films. Clear coats are essential to provide abrasion resistance and proper gloss level.

Some auto finishing shops apply vinyl spray paint on lower body panels. This abrasion-resistant coating protects these panels from being chipped by stones and other flying objects.

Flex (Elastomeric) Additives. Flex additives import flexibility to most topcoat qualities. They can be blended with acrylic lacquers, acrylic enamels, polyurethane acrylic enamels, and acrylic urethanes to produce flexibility for bumpers, fascia, and other flexible parts. The basic characteristics of the topcoat remain the same except for the flexibility.

Fisheye Eliminators. Fisheye eliminators are added to acrylic enamels, acrylic enamel base coat/clear coat, synthetic enamels, and urethane enamels to keep fisheyes under control. All additives must be used according to the manufacturer's instructions on the label. They make the paint rough, avoid use if at all possible.

Antichipping Paint. Most modern automobiles come with an antichipping paint applied to the rocker panels. Compared with other paints, antichipping paint has better coherence and pliability and helps prevent rust occurrence due to stone damage. When repainting, care is needed to apply antichipping paint at the designated areas so as not to lower the anticorrosion effectiveness.

Rocker Black Paint. This is either a semigloss or flat enamel coat that is applied to the rocker panel. After first masking off the surrounding area to protect it from paint overspray, the rocker panel is hand sprayed with a two-component urethane paint that has good adhesion qualities. If there are any other panels, such as the back panel, that require blacking, they should also be sprayed.

REVIEW QUESTIONS

1. Technician A says that a hardener added to enamel improves gloss. Technician B says that a hardener added to enamel slows the drying time. Who is right?
   a. technician A
   b. technician B
   c. both technician A and technician B
   d. neither technician A nor technician B
2. Lacquer paints dry by _____ of the solvent.
3. Technician A says that enamel paint is more durable than lacquer paint. Technician B says that dried lacquer paint will turn into a liquid if it encounters thinner. Who is right?
   a. technician A
   b. technician B
   c. both technician A and technician B
   d. neither technician A nor technician B
4. Paint is made up of
   a. binder.
   b. pigment.
   c. solvent.
   d. all of the above.
5. Technician A uses a primer over bare metal. Technician B uses a sealer over bare metal. Who is right?
   a. technician A
   b. technician B
   c. both technician A and technician B
   d. neither technician A nor technician B
6. Most of the paint work done by a paint shop is panel repair.
   a. True
   b. False
7. Technician A says that some epoxy primers require an induction time. Technician B says that self-etching primer is also known as wash primer. Who is right?
   a. technician A
   b. technician B
   c. both technician A and technician B
   d. neither technician A nor technician B
8. _____ are sprayed to fill scratches.
9. Technician A says that acrylic lacquers do not meet VOC requirements. Technician B says that adding hardener to enamel gives it a pot life. Who is right?
   a. technician A
   b. technician B
   c. both technician A and technician B
   d. neither technician A nor technician B
10. Cold, humid weather produces the fastest paint dry time.
    a. True
    b. False

Chapter

8

Sanding Materials and Methods

Objectives

After reading this chapter, you should be able to:
- Select the correct abrasive and sanding techniques for specific sanding operations.

Key Term List

grit
high crown
low crown
bullseye

The life of a finish and the appearance of that finish will depend considerably upon the condition of the surface over which the paint is applied. In other words, proper surface preparation is the foundation of a good paint job. Without it, there may be a weak base for the topcoat that eventually can result in the failure of the finish. This chapter will explain one important aspect of surface preparation, proper sanding methods.

8.1 — SANDPAPER

When sandpaper is constructed, a flexible or semi-rigid backing attaches to the abrasive grains, which are bonded by an adhesive. Hence, the most efficient results for a particular application depend on the selection and manufacturing of suitable combinations of grains, adhesives, and backings available. The automotive refinishing professional must then select and correctly use the proper sandpaper product for optimum productivity, material cost efficiency, and the best finish.

ABRASIVE TYPES

The abrasive grains or **grits** used to manufacture sandpaper products for automotive refinishing are selected on the basis of their hardness, toughness, resistance to grinding heat, fracture characteristics, and particle shape. The kind of grain a refinisher chooses depends on the purpose for which the coated abrasive is to be used.

As for abrasive types, most paint shops stock two: silicon carbide and aluminum oxide. Silicon carbide is a very sharp and fast-penetrating grain, customarily used (in paper sheet and disc form) for featheredging and dry sanding soft materials, such as old paint, fiberglass, and body putty. The major limitation of silicon carbide grain is that it tends to break down and dull rather readily when sanding hard surfaces.

Aluminum oxide is an extremely tough, wedge-shaped grain that better resists fracturing and dulling. Traditionally popular in coarse grits for grinding damaged metal, stripping old paint, and shaping plastic filler, numerous tests have demonstrated the superior performance of aluminum oxide sanding sheets and discs over silicon carbide on today's modern paint systems. Aluminum oxide is also preferred for use with today's paint finishes, which are predominantly basecoat/clear coat, have harder surfaces, and are applied in thinner layers than traditional lacquers and enamels. The blocky shape of the aluminum oxide abrasive when compared to silicon carbide makes it less likely to create deep scratches right through to the base material and so reduces the risk of overcutting. The greater durability of aluminum oxide versus silicon carbide enables the abrasive sheet or disc to better resist edge wear and dulling for longer effective life on these harder finishes.

A third type of abrasive, zirconia alumina, has been developed through advanced technology and continues to gain widespread preference in paint shops. Zirconia alumina grain has a unique, self-sharpening characteristic that provides continuous new cutting points during the sanding operation for reduced labor, increased efficiency, and longer effective life compared to traditional abrasives. Also, the fact that zirconia alumina products run cooler is particularly important when removing OEM clear coat finishes because of the extra heat generated when sanding these harder paint surfaces. A hot-running disc or sheet will load faster as the material being sanded softens and "balls up" in the abrasive. The self-sharpening action reduces the amount of sanding pressure required. Often refinishing professionals find that they can save money by using one grit finer and getting a better finish. The net result is that zirconia alumina abrasive products are being recognized as the more cost-effective alternative to traditional aluminum oxide and silicon carbide for a growing number of auto body repair and refinish operations.

GRIT NUMBERING SYSTEM

The rough side of the sandpaper is called the grit side. Grit sizes vary from coarse to micro fine grades and are ordered by number (Table 8–1). The lower the number, the coarser the grit. For example, a 24 grit is used to remove old paint film, while a 320, 360, or 400 grit is used to sand the gloss of an old finish to be repainted. Very fine and ultrafine abrasive papers are used primarily for color coat sanding. The so-called compounding papers, the 1,250, 1,500, and 2,000 grits, are used to solve problems on basecoat/clear coat paint surfaces such as those shown in Figure 8–1.

TABLE 8–1:   TYPES OF GRIT AND NUMBERING SYSTEM

Grit	Aluminum Oxide	Silicon Carbide	Zirconia Alumina	Primary Use for Auto Body Repair
Microfine	—	2000 1500 1250	—	Used to correct clear coat defects.
Ultrafine	—	1000	—	Used for clear coat sanding.
Very fine	— 400 320	600 400 320	600 400 —	Used for sanding surfacer and old paint prior to painting.
	220	220	—	Used for sanding of existing OEM paint.
Fine	180 150	180 150	180 150	Used for final sanding of bare metal and filler.
Medium	120 100 80	120 100 80	— 100 80	Used for smoothing old paint and plastic filler.
Coarse	60 50 40 36	60 50 40 36	60 — 40 —	Used for first sanding of plastic filler.
Very coarse	24	24	24	Used on sander or grinder to remove paint.

All domestic manufacturers conform to the same grading system for uniform consistency of standards. Differences in performance when using the same mineral, grit, bond, and backing from different manufacturers can be attributed to differences in manufacturing processes or quality, and/or operator methods.

As shown in Figure 8–2, the abrasive papers are available in various sizes and shapes. The most common forms found in paint shops are sheet stock and discs. The sheet stock—usually nine by eleven

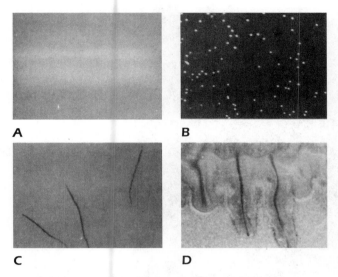

FIGURE 8–1 Ultrafine sandpaper can be used to remove (A) orange peel, (B) dirt, (C) small scratches, and (D) runs.

FIGURE 8–3 Applying a sandpaper disc to a DA.

FIGURE 8–2 Various sizes and shapes of abrasive papers and discs.

FIGURE 8–4 Velcro-backed paper and board.

inches—can then be cut into smaller pieces. Sheets are also available in jitterbug and board or body file sizes.

The most common abrasive sanding disc sizes for disc and dual-action sanders are five, six, and eight inches. Sandpaper disc grit sizes generally range from 50 to 400 grit. When using a self-adhesive sandpaper disc (Figure 8–3), be sure to center the paper on the pad before pressing it into place. Immediately after finishing the sanding operation, remove the used sandpaper from the backing pad. If it is not removed

right away, the adhesive will harden and cause the disc to stick fast to the backing pad. Should this occur, use solvent on a rag to dissolve the adhesive and then remove the paper. Some abrasive disc pads have a velcro backing to be used with a velcro pad. These allow easy disc changing and reuse (Figure 8–4). Sand papers for use with a vacuum system have holes to pickup dust (Figure 8–5).

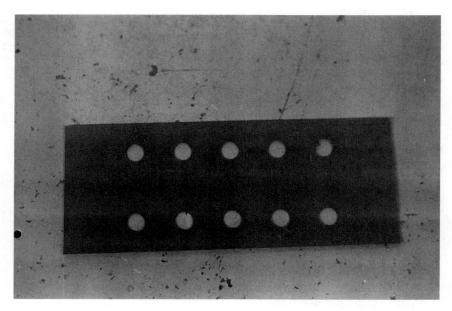

FIGURE 8–5   Vacuum holes in the sandpaper.

FIGURE 8–6   Grinder.

Grinding abrasive discs are used for rough jobs, such as grinding off rust and paint. They are available in numbers from 24 to 50 grit and in diameters of three to nine inches. A dual-action sander can be used to remove light rust, but heavy surface rust must be removed with a grinder (Figure 8–6). The grinder disc is first assembled to the backing plate, and then the disc/plate assembly is attached to the grinder (Figure 8–7). Some sandpaper discs are available with a center hole and are fastened to the sander in the same manner as the grinding abrasive disc. This manner of fastening is necessary in some wet-sanding operations.

Although grinding discs are thicker and stronger than sandpaper discs, they are rather thin and easily bent. For this reason, the backing plate is necessary

A

B

FIGURE 8–7 *(A) Grinder parts. (B) Assembling grinder parts.*

to provide stiffness for the revolving disc. They are available in different sizes. After use, usually only the outer edge of a grinder disk is worn. A disk trimmer can be used to cut the worn edge of the disk off (Figure 8–8). The smaller disk is used with the proper-sized backing plate.

COATED ABRASIVE SURFACES

Coated abrasives are generally manufactured in two types of surface distributions (Figure 8–9):

- Closed-coat abrasive paper
- Open-coat abrasive paper

FIGURE 8–8 *Trimming a worn disc.*

FIGURE 8–9 *Closed-coat (left) and open-coat (right) sandpaper.*

A closed-coat product is one in which the surface grains completely cover the sanding side of the backing. An open-coat product is one in which the abrasive grains are spaced to cover between 50 and 70 percent of the backing surface.

As for uses, open-coat products are the popular choice on softer materials such as old paint, body filler, and putty, plastic, and aluminum, where premature loading of the abrasive would otherwise be a problem. Closed-coat products generally provide a finer finish and are most commonly used in wet-sanding applications.

In addition to open-coat construction, many abrasive sheets and discs are surface-coated in manufacturing with a zinc stearate solution to further prevent the premature loading of the sandpaper and to extend its useful life. This is particularly true of fine-grit papers commonly used for scuff sanding old paint and surfacer and finishing body filler. During those applications, the materials tend to soften because heat is generated while sanding and loading the abrasive. Remember also that with talc-coated products, the coating breaks away from the abrasive during use, taking with it sanding residue and thereby freeing the abrasive to cut longer.

Talc remains an excellent load-resistant feature. However, with the evolution of paints, in particular today's popular basecoat/clear coat paint systems, certain elements of surface preparation become more critical, necessitating a possible alternative to the standard zinc stearate–coated sanding product.

One critical element is a contamination-free surface. Although talc can extend abrasive life, it also contributes another contaminant to the auto body surface. A clean surface under the basecoat is extremely critical for the success of the basecoat/clear coat paint job—more so than with standard enamels and lacquer. In the past, dust "nibs" or contamination under the first color coat could be sanded down, spot recoated, and blended in without detracting from the appearance of the final finish. With basecoat/clear coat, however, any defects are magnified by the clear coat. With metallic flake and pearl color coats, subsurface dirt is a disaster. A key, then, is to minimize the contaminants on the auto body surface whenever possible. To this end, new sanding products have been introduced for auto body and paint professionals that employ high-tech, antistatic bonding agents to retard loading without zinc stearate.

In addition, new advanced abrasives such as zirconia alumina, anchored by resin-type adhesives, are becoming more and more popular in the automotive refinishing trade because they cut faster and cooler. In other words, they will cut through paints, primers, and plastic fillers before they can soften and load the abrasive. And from a safety standpoint, sanding with products that do not have a zinc stearate coating creates less nuisance dust in the air as well as on the auto body surface.

WEIGHT OF PAPER

The proper selection of backings likewise depends on the application involved. Paper-backed abrasive products used in automotive refinishing are designated under uniform standards by all manufacturers as A-, C-, D-, or E-weight. A-weight paper is the lightest, most conformable paper backing available. It is popular for wet color sanding and dry finish sanding. The C- and D-weight paper products are progressively heavier, tougher, and less flexible. They are suitable for coarser sanding applications. E-weight paper is being more widely used by refinishing personnel for paint stripping and shaping of filled areas, as it is more durable than the traditional D-weight paper backings once popular for these applications. D- and E-weight papers are sometimes referred to as _production papers_ because their construction produces a fast-cutting, long-lasting abrasive surface.

Cloth backings employed in products used by the auto body trade are likewise designated by a letter code. J-weight is a light, flexible cloth, popular for general clean-up and deburring in sheet or handy roll form. X- and Y-weight cloths are heavier, more rigid backings often used in small disc form for tight-quarter coarse sanding.

Fiber backings are most common in grinding discs. This very tough, semirigid backing is best suited for heavy operator pressure applications, such as weld grinding and rust removal. The most suitable fiber backing for automotive application is 30 mil vulcanized fiber because of its extra durability and greater resistance to breakdown and edge chipping.

SAFETY POINTERS WITH ABRASIVES

The following points must be kept in mind when working with abrasives:

- Grinding discs should never be run if the edges are nicked, torn, or show excessive wear. Whenever in doubt, do not use the product. Recommended fiber disc grinding speeds are 5 inch, 7,650 rpm; 7 inch, 5,500 rpm; 9-1/8 inch, 4,200 rpm.
- Fiber grinding discs should be seated flat against a back-up pad and should never overhang a pad by more than one-quarter inch.
- When paper discs are used on a slow-speed polisher, the recommended speed is 3,000 rpm or less.
- Curled discs generally indicate improper storage and should not be used until the shape is corrected. Storage of discs at 65 to 75 °F will prevent excessive curling of abrasive products prior to usage.

8.2 METHODS OF SANDING

Refinishing sanding can be done:

- by hand or
- by power equipment.

Most heavy sanding, such as sanding old finish, is done by power sanders. But some conditions, particularly the delicate operations, dictate hand-sanding.

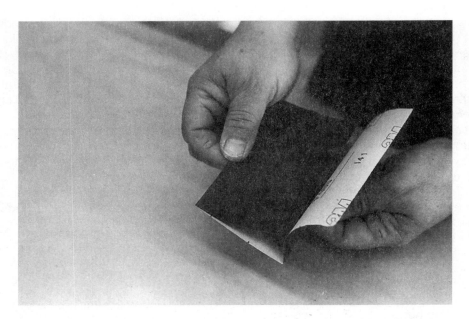

FIGURE 8–10   Fold sandpaper into thirds.

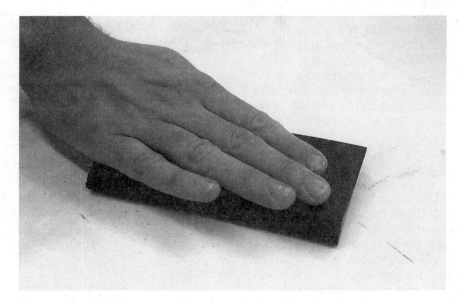

FIGURE 8–11   Hand sanding.

HAND-SANDING

Hand-sanding is a simple back and forth scrubbing action with the sandpaper flat against the surface. It can be achieved by following a general procedure such as this:

1. Cut the sheet of sandpaper in half crosswise and then fold in thirds (Figure 8–10).

2. Place the paper in the palm of the hand and hold it flat against the surface. Apply even, moderate pressure along the length of the sandpaper using the palm and extended fingers (Figure 8–11). Sand back and forth with long, straight strokes. If the palm of the hand is not flat on the surface, the fingers will be doing the sanding. This will result in uneven pressure

1 After analyzing the dent and planning the repair, the paint is removed. The crease is visible near the center of the bare metal area.

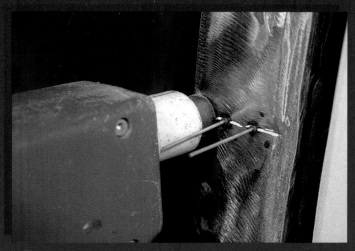

2 Draw pins are welded to the crease.

3 A slide hammer is attached to the pins and the crease is pulled out.

4 The pins are ground off after the pulling is complete.

5 Plastic body filler is spread over the remaining crease.

6 A cheese grater is used to quickly level the semi-cured plastic filler. When the filler has thoroughly cured, it is sanded with 40, 80, and 180 grit paper.

7 Plastic protects the vehicle from overspray. Masking paper is applied adjacent to the spray area.

8 Three-part surfacer is mixed in a separate container. It is filtered as it is poured into the paint cup.

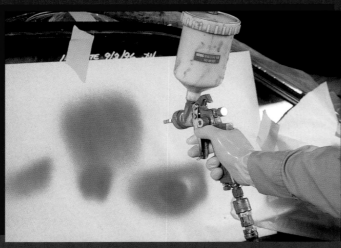

9 The gravity feed paint gun is spray-tested on the masking paper before spraying the vehicle.

10 Two or three coats of surfacer will fill 180 grit sandpaper scratches.

11 A light dusting of contrasting color paint is sprayed over the surfacer. This guide coat will aid in block sanding.

12 The surfacer is block sanded. Guide coat remaining after block sanding indicates low spots or scratches. Bare metal indicates high spots.

13 The contaminated masking materials are removed during prep for topcoat.

14 A blow gun is used to remove dust and debris from all crevices.

15 Liquid mask is sprayed on the engine. The liquid mask dries to a plastic-like film that protects surfaces and holds dirt. It is removed with water.

16 Final masking before topcoat is finished in the spray booth.

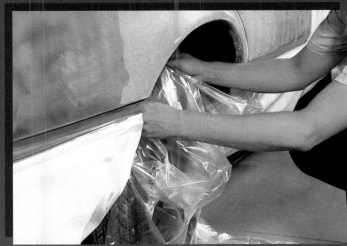

17 Tires can be protected from overspray with plastic wheel masks.

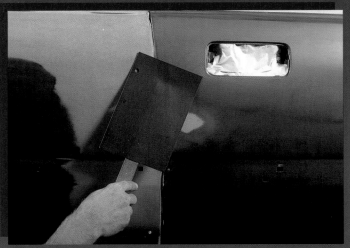

18 A mismatch between a test panel and the vehicle is shown.

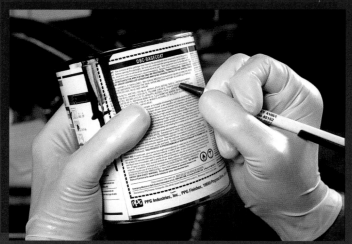

19 Read the paint can label for safety, mixing, and application information.

20 A digital scale is used to weigh the required amount of tint to make paint.

21 A paint stick is used to measure the proper amount of base coat and reducer.

22 The mixed paint must be filtered as it is poured into the paint gun cup.

23 This replacement hood is tacked off immediately before sealing.

24 A black sealer is applied to the replacement hood prior to base coat.

25 This HVLP gravity feed spray is used to apply the base coat. It is adjusted prior to use.

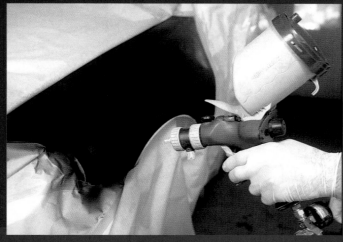

26 Base coat application on a fender. The gun is always held perpendicular to the surface.

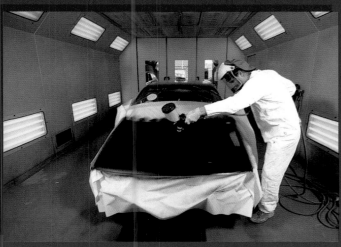

27 Clear coat is sprayed on the base coat. Notice the gloss in the clear coat. Start at one side and work towards the opposite side.

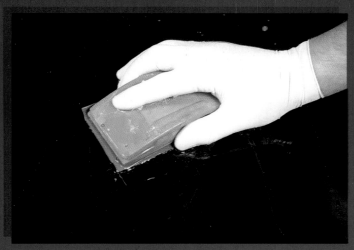

28 Fine sandpaper (1500-2000 grit) will remove clear coat imperfections.

29 The sand scratches are removed by buffing. In this case, a foam pad is used.

30 Detailing of the vehicle prior to delivery includes a thorough washing.

PAINT FAULTS

Dust in finish

Runs

Blistering

Mottling

Sand scratch swelling

Fisheyes

Cold cracking

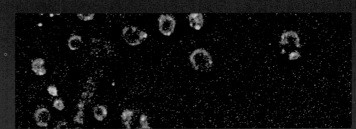

Burned through

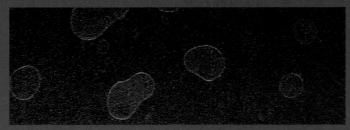

Water spotting

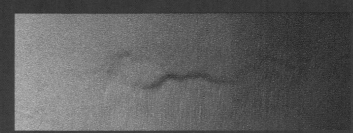

Metallic sag

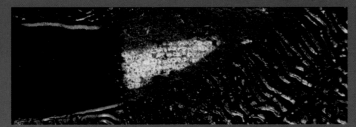

Lifting

Chemical spotting

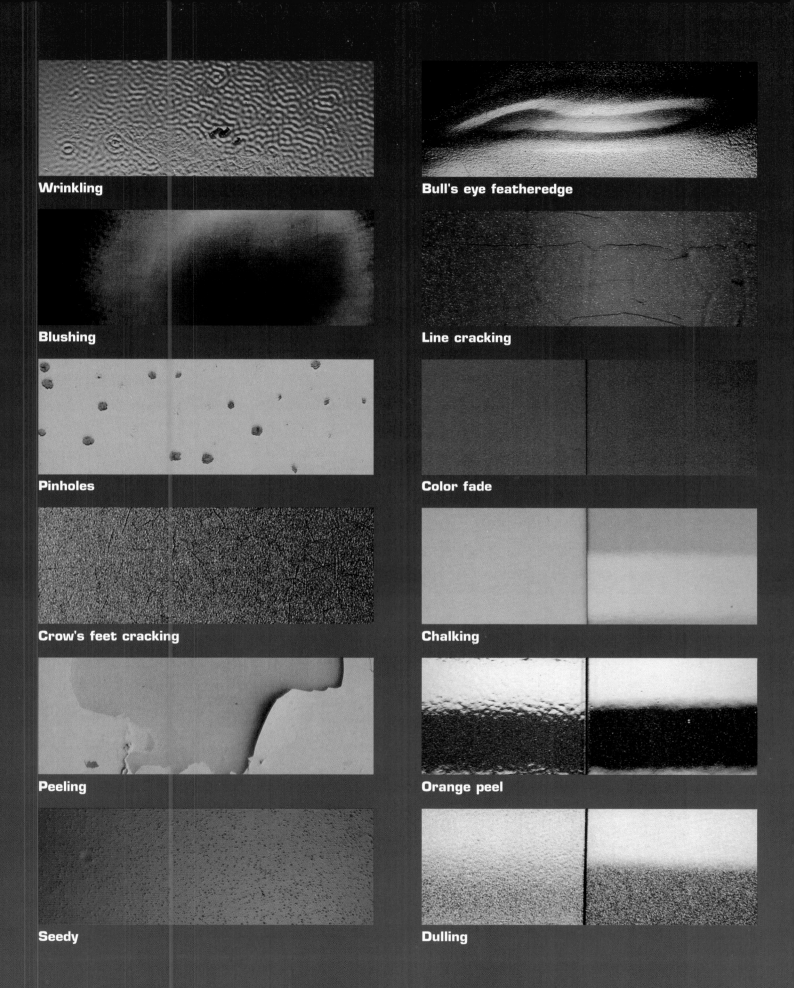

Wrinkling

Bull's eye featheredge

Blushing

Line cracking

Pinholes

Color fade

Crow's feet cracking

Chalking

Peeling

Orange peel

Seedy

Dulling

TINTING

BASE COLOR	ALUMINUM LETDOWN	MASS TONE	WHITE LETDOWN	BASE COLOR	ALUMINUM LETDOWN	MASS TONE	WHITE LETDOWN
1. Yellow Gold				15. Indo Orange			
2. Lt. Chrome Yellow	Not to be used with Aluminum Letdown.			16. Moly Orange (Red Shade)	Not to be used with Aluminum Letdown.		
3. Oxide Yellow				17. Red Oxide			
4. Indo Yellow				18. Transparent Red Oxide			
5. Transparent Yellow Oxide				19. Deep Violet			
6. Rich Brown				20. Quindo Violet			
7. Black				21. Magenta Maroon			
8. Strong Black				22. Phthalo Green (Yellow Shade)			
9. Organic Orange (Light)				23. Phthalo Green			
10. Oxide Red				24. Scarlet Red	Not to be used with Aluminum Letdown.		
11. Permanent Red				25. Perrindo Maroon			
12. Organic Scarlet				26. Phthalo Blue (Medium)			
13. Phthalo Blue (Green Shade)				27. Phthalo Green			
14. Permanent Blue				28. Phthalo Green (Yellow)			

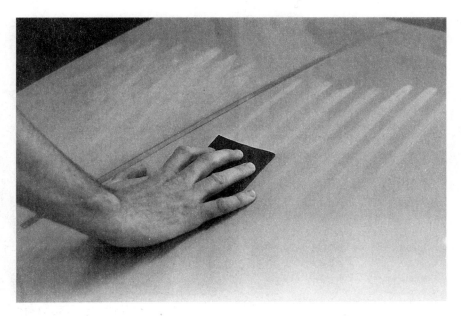

FIGURE 8–12   Finger sanding.

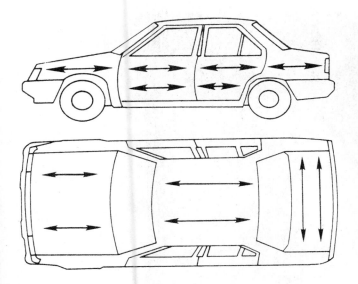

FIGURE 8–13   Sand with the body lines.

FIGURE 8–14   Abrasive pad used in hard-to-reach areas.

being applied in the spaces between the fingers (Figure 8–12). Finger sanding should be avoided.

3. Do not sand in a circular motion. This will create sand scratches that might be visible under the paint finish. To achieve the best results, always sand in the same direction as the body lines on the vehicle (Figure 8–13).

4. Be sure to thoroughly sand areas where a heavy wax buildup can be a problem, such as around trim, moulding, door handles (Figure 8–14), and radio antennae, and behind the bumpers. An abrasive pad will easily conform to the curved contour. Paint will not adhere properly to a waxy surface.

FIGURE 8–15   *Sanding pad.*

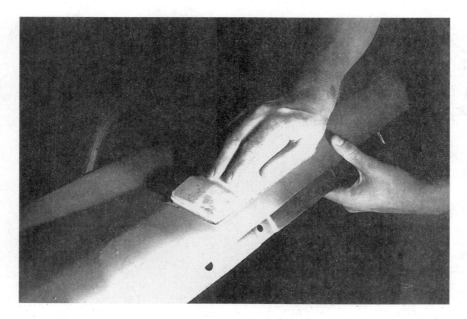

FIGURE 8–16   *Sanding block.*

5. Use a sanding block or pad for best results. To sand convex or concave panels (Figure 8–15), employ a flexible sponge rubber backing pad. Use a sanding block (Figure 8–16) to sand level surfaces.

6. When hand-sanding primer or putty, make certain to sand the area until it feels smooth and level. Rub a hand or a clean cloth (Figure 8–17) over the surface to check for rough spots.

Hand sanding may be done dry or wet.

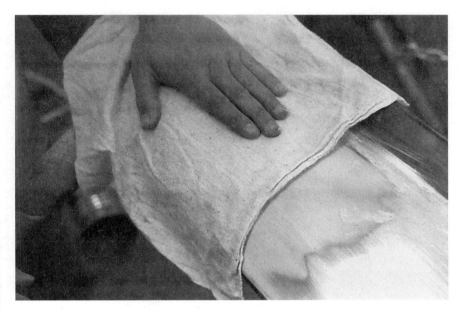

FIGURE 8–17   *A cloth in between the hand and the surface makes it easier to feel problems.*

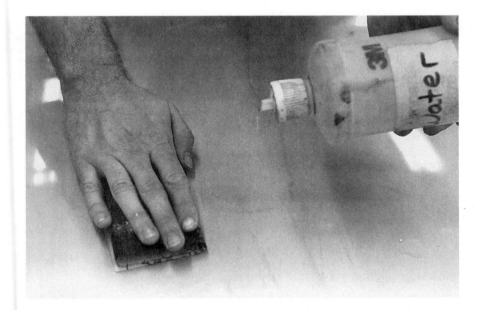

FIGURE 8–18   *Water bottle and sanding pad.*

Dry Sanding

This is basically the back-and-forth procedure just described. But one of the problems with it is that the paper tends to clog with paint or metal dust. Tapping the paper from time to time will remove some of the dust. Another suggestion is to use talc-coated or dry-lubricated sandpaper, which tends to prevent clogging. Specially treated open-coat paper resists loading for long life.

Wet Sanding

Wet sanding also solves the problem of paper clogging. It is basically the same action as dry sanding except that water, a sponge, and a squeegee are used in addition to the sanding block. Sandpapers are available in dry, wet, or wet-or-dry abrasive types.

When wet sanding, dip the paper in the water or wet the surface with a water bottle (Figure 8–18). Use plenty of water, employing short strokes and light

FIGURE 8–19 *A squeegee is used to remove water when wet sanding.*

pressure. Never allow the surface to dry during the wet-sanding operation. Also, do not allow paint residue to build up on the abrasive paper. It is possible to tell how well the paper is cutting by the amount of drag felt as it moves across the surface. When the paper begins to slide over the surface too quickly, it is no longer cutting. The grit has become filled with paint particles or sludge. Rinse the paper in water to remove the paint and sponge the surface to remove the remaining particles. Then the sandpaper will cut the surface again. Check the work periodically by sponging the surface off and wiping it dry with a squeegee. This will remove all excess water, so that it is easier to evaluate the surface condition. It is usually wise to complete one panel or body section at a time, then remove the sanding residues with the sponge and dry off with a squeegee before sanding the next panel (Figure 8–19).

Once the wet-sanding operation is completed, be sure that all surfaces are dry. Blow out the seams and mouldings with compressed air at a lower pressure and tack-rag the entire surface.

A comparison of the advantages and disadvantages of wet and dry sanding is given in Table 8–2.

POWER SANDING

As described in Chapter 2, there are four types of power sanders used by the refinisher (Figure 8–20):

TABLE 8–2:	COMPARISON OF WET AND DRY SANDING	
Item	**Wet Sanding**	**Dry Sanding**
Work speed	Slower	Faster
Amount of sand-paper required	Less	More
Condition of finish	Very good	Final finish difficult
Workability	Normal	Good
Dust	Little	Much
Facilities required	Water drain necessary	Dust collector and exhaust necessary
Drying time	Necessary	Not necessary

- Disc sander or grinder
- Orbital or jitterbug pad sander
- Dual-action (DA) sander.
- Straight line or board sander

All four types of sanders are powered by air.

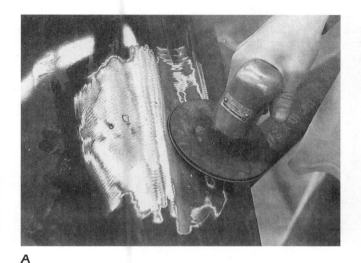

A

B

C

D

FIGURE 8–20 *(A) Grinder. (B) Jitterbug. (C) DA. (D) Air file.*

In general, the type of power sander dictates sanding procedures (Table 8–3). Disc sanders or grinders, for example, have high-speed discs that turn from 2,000 to 6,000 rpm. They use circular discs from five to nine inches in diameter and are used for such operations as grinding off an old finish. Heavier grinders (Figure 8–21) generally take a nine-inch-diameter disc and—because of the obvious safety hazard involved—many have both a rear and a side handle for better control.

When using a disc grinder, one must tilt it slightly so that only about one inch of the leading edge of the sanding disc contacts the surface (Figure 8–22).

Never use the disc flat on the surface, because it twists the grinder and can even cause it to fly out of one's grip. Also, when held flat, it makes circular sand scratches, which are difficult to get rid of. Never use a disc grinder at a sharp angle with just the edge of the disc in contact, because this will cause it to gouge or dig deeply into the surface. When a disc grinder is properly held, the sanding marks are nearly straight.

Orbital sanders have an eccentric (off center) action that produces either a partly circular scrubbing action (orbital pad or dual-action types) or a straight back-and-forth reciprocating action (flat orbital or straight-line type). Unlike the disc sander just

TABLE 8–3:   USE OF SANDERS

Sander Type	Normal Area of Operation	Normal Use							
		Sanding old finishes	Paint stripping	Feather-edging	Rough sanding or solder	Rough sanding of metal putty	Rough sanding of poly putty	Sanding of metal putty	Sanding of poly putty
Disc Sander		A	A	C	B	A	C	C	C
Dual-Action Sander	Suitable for narrow areas	A	B	A	C	C	C	C	C
Orbital Sander		B	C	B	C	C	C	C	C
Straight-Line Sander	Suitable for wide open spaces	C	C	B	C	A	A	A	B
Long Orbital Sander		C	C	B	C	A	A	A	B

NOTE: It is important that the correct type of sander and abrasive paper be used for each type of job. Also, always wear a mask or use some sort of dust arrester when using the sander.
A Preferred
B Acceptable
C Least preferred

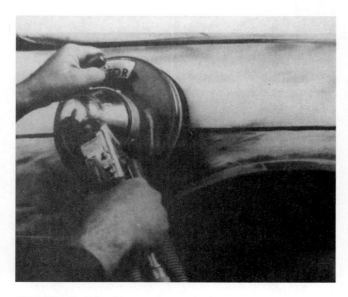

FIGURE 8–21   Heavy-duty grinder.

FIGURE 8–22   Only the upper inch of the grinding disc is in contact with the surface.

A

B

FIGURE 8–23   (A) Set air pressure at wall. (B) Hold DA flat.

discussed, orbital sanders should be pressed flat so they will not leave surface scratches. Orbital and board sanding can be done either dry or wet.

To operate an air sander, set the air pressure at sixty-five to seventy pounds. If you are right-handed, hold the handle of the sander in the right hand while using the left hand to apply light pressure and guide the tool (Figure 8–23).

To protect the chrome from damage, do not machine-sand closer than one-half inch from the trim and mouldings (Figure 8–24). Mask nearby trim, decals, glass, handles, and emblems to prevent metal

FIGURE 8–24 Stay at least one-half inch away from the chrome or other surfaces that should not be sanded.

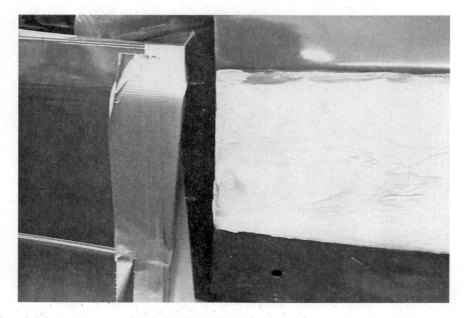

FIGURE 8–25 The right quarter panel is being repaired on this Camaro. The technician has loosened the rear bumper. He has added duct tape for extra protection.

sparks from pitting these surfaces. In fact, it is a good idea to remove mouldings and trim before sanding. If this is not possible protect trim and mouldings with cardboard and duct tape (Figure 8–25).

When using any mechanical sander—and particularly a disc grinder—keep it moving so that no deep scratches, gouges, or burn-throughs develop. And do not, except when sanding bare metal, power-sand styling lines, as this will quickly distort the styling edge.

WARNING: Always wear a dust mask (Figure 8–26A) when sanding, and wear both a dust mask and a face shield (Figure 8–26B) and hearing protection when grinding.

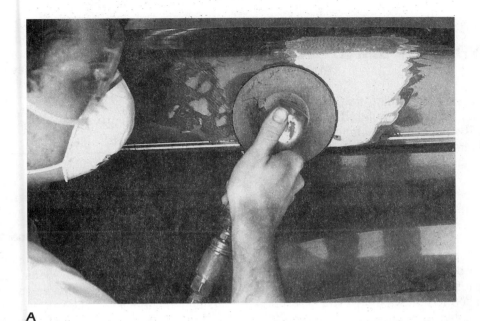

A

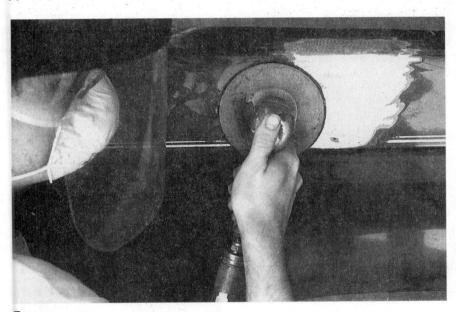

B

FIGURE 8-26   (A) Dust mask when sanding. (B) Dust mask and face shield when grinding.

When power sanding, replace the sandpaper when paint begins to cake or "ball" up. This paint buildup can scratch the surface and reduce the sanding action of the disc. Slowing down the speed of the sander will also help to prevent paint buildup on the sanding disc and to prolong sandpaper life.

SHOP TALK

The sandpaper grits listed in the text are advisory. Consult the paint manufacturers' manual for recommended sandpaper grits.

8.3 — TYPES OF SANDING

There are several types of sanding that a refinisher must master. Some can be completed with power sanders alone, others with a combination of power and hand, and still others by hand alone.

GRINDING

Grinding with a 24-grit disc is used to remove paint and/or metal. Always begin with a fresh disc because as the disc wears, it will produce more heat, possibly enough to warp the panel. Hold the grinder so that only the upper inch of the disk is in contact with the panel. The buffing technique—making only one pass over the area—is used to remove paint. The crosscut technique—making two or more passes over the area—is used to remove metal (Figure 8–27). Some DA sanders have a shaft lock that will convert the random orbital action to simple rotary action. With the shaft locked, the DA can be used as a grinder. When equipped with 80-grit paper, the DA can be used to

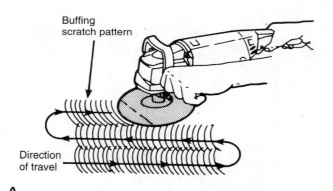

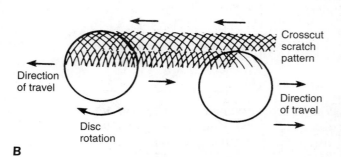

FIGURE 8–27 (A) Buffing removes paint. (B) Crosscutting removes paint and metal.

quickly remove paint. Hold the DA at a low angle, not flat, when using it as a grinder. It does not produce as much heat as a 24-grit disc grinder (Figure 8–28). Also, it does not leave deep scratches in the metal.

WARNING: Be sure to wear eye protection, ear protection and a dust mask. Leather gloves may be worn to protect hands.

Safety Pointers when Grinding

In addition to the safety procedures given in Chapter 2, the following pointers must be remembered:

- When disc grinding, hold the grinder firmly at a five-degree angle to the work surface.
- Grinding should direct sparks and dust away from the face and, if possible, toward the floor. Sparks from grinding may damage glass. Protect glass from spark pitting by covering the glass with a welding blanket.
- Be conscious of the grinder or polisher cord at all times to prevent entanglement.
- Do not grind or sand too close to trim, bumpers, or any projection that might snag or catch the grinding disc's edge.
- Start or stop a disc grinder in contact with the work surface.
- Never "free run" a grinding disc or set a grinder down until it stops completely.
- Make certain the backup pads are designed for the work and are free of cuts or nicks at the edge or at the center hole. Make certain that pads are seated on the shaft properly. Check for proper balance. Retainer nuts should not show

FIGURE 8–28   Hold the DA at a low angle.

excessive thread wear, must have at least three-thread contact, and should not cause damage to the grinding discs.

- Backup pads for use with the self-adhesive-type discs must be dry, clean, and dust free. Avoid using pads with a frayed, torn, dirty, or paper-contaminated surface. If necessary, wipe the pad face with a clean, dry cloth.

STRIPPING

Stripping is removing all of the paint on a panel. Several types of sanders can be used to do this. The biggest and fastest is a buffer equipped with a seven-inch, 24-grit disc. The operation is similar to grinding. The buffer is held at a slight angle (Figure 8–29). Always keep the buffer moving. Do not strip in any one area more than one minute. Keep moving from area to area to prevent heat buildup. Hoods, roofs, and decklids are **low-crown** panels that can be easily warped by heat from careless stripping. If the paint loads up the disc quickly, probably because it is lacquer paint, then you should consider chemical stripping, Chapter 9 explains chemical stripping. A grinder can be used to strip paint. Again, be alert to heat buildup. DAs and jitterbugs with 80-grit paper are useful for removing an OEM finish.

AIR FILING

Air filing is used to rapidly level body filler. Start with 40-grit paper on an air file (Figure 8–30). The file should span the filled area so that the ends rest on the undamaged metal. If the surface is a low crown, the air file can be held flat. On high-crown surfaces, the file should be turned on edge so only part of the paper is in contact with the filler (Figure 8–31). Stroke the file up and down with the crown. When most of the filler is sanded, switch to 80-grit paper to remove the deep 40-grit scratches. Finish with 180-grit paper.

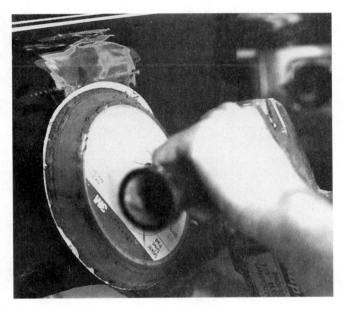

FIGURE 8–29   Buffer equipped with a 24-grit disc should be held at a low angle to quickly strip the paint.

FIGURE 8–30   On a low-crown surface, the air file can be held flat.

FIGURE 8–31   On a high-crown area, only the edge of the air file should be in contact with the surface.

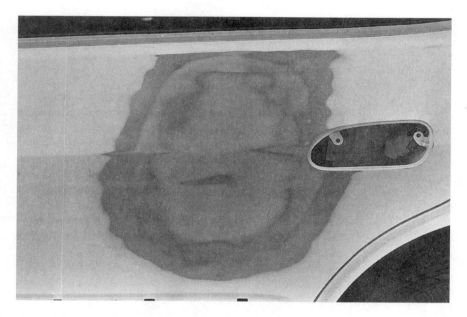

FIGURE 8–32   Featheredging.

FEATHEREDGING

Featheredging is the first step in repair of a broken paint edge. After featheredging, the repair area is sprayed with primer then surfacer. After the surfacer cures, the repair area is block-sanded. If each step is performed correctly, the repair will be invisible under the topcoat.

Featheredging can be done with power sanders or by hand. If you are using a power sander, a jitterbug or a finish DA are the best choices. A hard block is a necessity for hand feathering. The reasoning behind featheredging is to make a gradual change in height, from the outermost to the innermost layers of paint. Most OEM finishes will have, at most, four layers. Ideal featheredging of this would have at least one inch of each paint layer showing (Figure 8–32). If, due to space constraints, one inch of each paint layer showing is not possible, one-half inch of each paint layer is adequate. In the case of a narrow featheredge after priming and surfacing, the surfacer must be carefully block-sanded. Improper block-sanding or surfacer shrinkage will create a bulls eye at the nar-

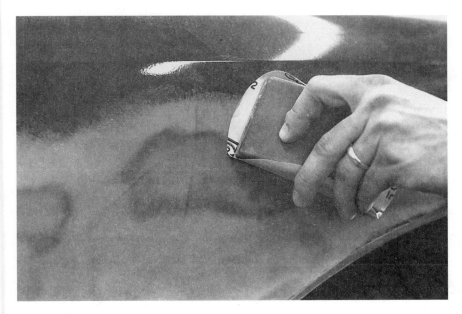

FIGURE 8-33   Hand feathering.

FIGURE 8-34   The door moulding below the bare metal has been taped. The moulding should be removed. There is not enough room to properly featheredge.

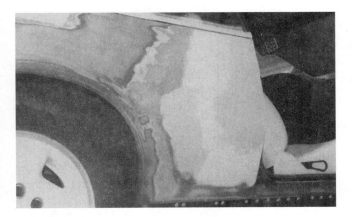

FIGURE 8-35   The two areas of bare metal should be consolidated to eliminate the island of paint between them.

row featheredge. If you are feathering a scratch or broken paint from a dent repair, the procedure is the same.

Start with 80-grit paper. Keep the sander flat at all times. Sand along the scratch or broken edge until there is at least one inch of each layer showing (Figure 8-33). Switch to 180- or 220-grit paper. Sand over the layers to remove the 80-grit scratches. Finish with 320- or 400-grit paper to achieve proper featheredging.

Here are some tips about featheredging:

- Do not try to featheredge an area where there is not enough room. An example would be if there was a scratch two inches below a stripe. Remove the stripe. Do not try to featheredge up to the stripe. There is not enough room for a proper featheredge.

- If you are working on a car that has a refinish on top of OEM paint so that there are six or more layers of paint, or about 12 mils, remove the refinish paint from the panel.

- Remove trim that may be in the way as shown in Figure 8-34.

- If you have two featheredged areas within four inches of each other, consolidate them (Figure 8-35).

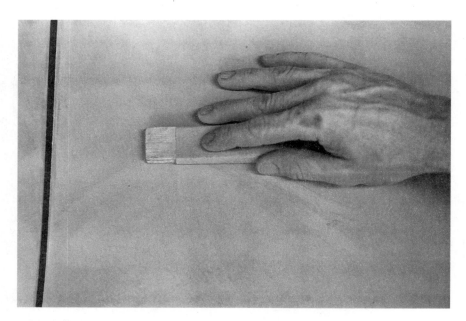

FIGURE 8–36 *A flat stick wrapped with sandpaper is an effective block sander.*

BLOCK SANDING

Featheredging and block sanding, when done correctly, will give the topcoat a flat base. Improper featheredging or block sanding will show in the top coat as a bulls eye. Because a bulls eye is slightly lower than the surrounding paint, it reflects light differently. The bulls eye may be visible as an edge or a circle. A bulls eye is a sign of poor surface preparation.

Block sanding is used to level surfacer. A guide coat, a light dusting of contrasting color, should be sprayed on the surfacer before block sanding. Block sanding may be done wet or dry. There are many methods, but the following has proved to be quite reliable: A flat, wooden paint mixing stick or a flat wooden ruler is wrapped with 320-grit dry sandpaper (Figure 8–36). The advantage of the wooden stick is that it is hard. The compression of the foam pad on hand blocks leads to a visible ring around the repair. The flat stick avoids this problem. Select the flat stick carefully; it must be completely flat. Anything other than completely flat will cause problems. Discard the stick if it becomes warped.

On small repairs, the stick should span the surfacer so that the ends rest on unsanded paint. On large repairs this will not be possible. Hold the stick flat on low-crowned surfaces. On **high-crowned** surfaces, hold the stick on edge. Sand with the crown of the panel. Then sand until the guide coat is removed. Low areas are indicated by guide coat. Span the area and sand the guide coat off. Do not dig into the low spot.

High areas are indicated by bare metal. If bare metal shows around it, you must go back to the metal finishing step. If there aren't any high areas and the guide coat is completely removed, you can blow off the panel and wet sand with 600-grit paper. Because the surface is now flat, you can use either a rubber block or hard pad. All you need to do now is sand out the 320-grit scratches.

SCUFFING

Scuffing means to use a red abrasive pad to take the shine off a new replacement part primer (Figure 8–37). The normal procedure for a fender is to scuff the outside of the panel as well as any inside areas that have OEM color. On hoods, decklids and doors, the inside area is scuffed. The purpose of scuffing is to abrade the primer enough to allow the sealer to bite into it. Scuffing is done by hand or with a pad attached to a velcro disk. As you scuff the replacement part, look out for runs in the factory primer. If these are not sanded out, they will show through the paint.

SANDING

Sanding means to use 320- or 400-grit sandpaper to roughen up the paint before using a sealer. All surfaces must be thoroughly sanded for the sealer to adhere. Paint will not stick to a poorly sanded surface. Sanding can be done by hand or with power equip-

FIGURE 8–37 This replacement door is scuffed with an abrasive pad prior to trimming.

A

B

ment. Always keep the sander or block flat to prevent gouges. Sand until the gloss is removed. Wet sanding cuts faster than dry sanding. Also, wet sanding produces large quantities of water on the floor, and dry sanding produces clouds of dust. As you wet-sand, use a squeegee to remove the water so that you can see the surface.

CLEAR SANDING

The clear coat on a blend panel needs to be sanded to accept the refinish clear coat. There are three methods of clear sanding. The fastest is to sand with a DA and a soft pad. Use 800-grit dry paper (Figure 8–38A). The soft pad will allow the DA to float over the body lines and will prevent the DA from sanding into the basecoat. Another method is to use 600- or 1,000-grit wet paper (Figure 8–38B). Sand with the body lines to minimize visible scratches. You only need to lightly sand the surface. Do not sand through the clear into the basecoat. The last method is to use a gray abrasive pad. Sanding paste can be used to make the pad cut faster (Figure 8–38C).

COLOR SANDING

Buffing time and effort can be reduced if the surface is sanded prior to buffing. Urethane clear coat has a window of time when it may be easily buffed. The clear coat needs enough time to partially cure, but not enough time to completely cure. Completely cured urethane clear coat can be buffed, but the surface is hard and the buffing is difficult. The window of time varies according to the type of clear coat and the

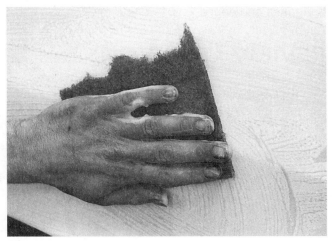

C

FIGURE 8–38 Sanding clear for a blend: (A) 800-grit dry paper on a DA, (B) 1,000-grit wet paper on a sanding pad, (C) gray scuff pad and sanding paste.

FIGURE 8–39 *Color sanding with 1,500-grit sandpaper prior to buffing.*

drying conditions. Check the manufacturers' recommendations for the proper time to buff. Fine grades of wet sandpaper (1,200, 1,500, and 2,000) can be used (Figure 8–39). Soften the sand paper prior to use by submerging it in water for twenty minutes. Use a block or hand pad. To remove heavy orange peel, use 1,200-grit paper followed with 1,500-grit paper. If you need to remove dirt, sand only on the dirt. Any scratches you put in will need to be buffed out. To remove light orange peel, use 2,000-grit paper. Dry color sanding is an option. In this method, use 1,500-grit sandpaper on a DA. Be very careful around the body lines and edges.

REVIEW QUESTIONS

1. Technician A says that the higher the grit number, the coarser the sandpaper. Technician B says that the lower the grit number, the finer the sandpaper. Who is right?
   a. technician A
   b. technician B
   c. both technician A and technician B
   d. neither technician A nor technician B

2. Which grit of sandpaper is used to sand the surfacer prior to painting?
   a. 80
   b. 400
   c. 1,000
   d. 1,500

3. Technician A uses the buffing technique when removing metal. Technician B uses the crosscut technique to remove paint. Who is right?
   a. technician A
   b. technician B
   c. both technician A and technician B
   d. neither technician A nor technician B

4. Block sanding should be done with hand-held sandpaper.
   a. True
   b. False

5. Technician A strips paint with a grinder. Technician B strips paint with a DA. Who is right?
   a. Technician A
   b. Technician B
   c. both technician A and technician B
   d. neither technician A nor technician B

6. Which weight of paper is the lightest?
   a. A
   b. C
   c. D
   d. E

7. Technician A hand-sands with the vehicle's body lines. Technician B says that dry sanding cuts faster than wet sanding. Who is right?
   a. technician A
   b. technician B
   c. both technician A and technician B
   d. neither technician A nor technician B

8. _____ sanding is used to level the surfacer.

9. Technician A color-sands with 600-grit sandpaper. Technician B color-sands with 1,500-grit sandpaper. Who is right?
   a. technician A
   b. technician B
   c. both technician A and technician B
   d. neither technician A nor technician B

10. An air file should be held flat on a high-crown surface.
    a. True
    b. False

Chapter 9

Surface Refinishing Prep

Objectives

After reading this chapter, you will be able to:
- Inspect the paint surface for defects.
- Remove obstructions.
- Properly clean the surface.
- Properly prep the surface.
- Mask the surface.
- Block-sand the surface.

Key Term List

blend
primer
surfacer
sealer
aggressive cleaner
mild cleaner
block-sand
metal conditioner
conversion coat
back tape
guide coat

Many collision repair facilities warrant repairs for as long as the owner has the vehicle. Because of this, surface prep is more important than ever. Each repair step builds on the previous step. Even a slick topcoat looks bad if it has bull's-eyes, pinholes, or sand scratches. Follow the steps listed in this chapter to prepare the vehicle surface for repainting.

9.1 — INSPECTION

Proper surface prep begins with a thorough evaluation of the surface. Paint defects are difficult to see in bright light. A paint surface is best evaluated inside a shop, out of direct sunlight. Wash the car if necessary. Search the surface for any of the following:

- Checking
- Rust
- Refinishing
- Peeling

Checking, line checking, crazing, cracking, or microchecking are indicators of serious problems. Often the paint will appear dull. If you look closely, you can see small cracks or "crow's feet" in the paint (Figure 9–1A & B). If any of these are present, the paint surface needs to be stripped to bare metal.

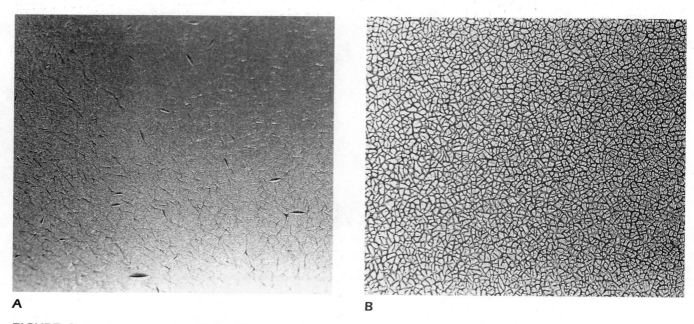

A   B

FIGURE 9–1   *Two examples of checking. In A, the surface looks dull. Close inspection reveals small crow's feet cracks in the paint. B is an extreme example.*

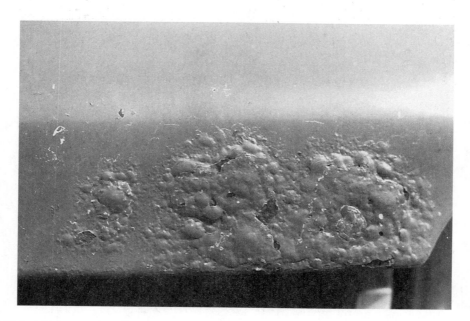

FIGURE 9–2   *Rust bubbles on a hood.*

Rust can range in appearance from obvious holes to small bubbles (Figure 9–2). Rust that begins on the inside and corrodes outward is often found on hem flanges (Figure 9–3). Hem flanges are used to hold an outer skin to a framework. These flanges are found on doors, hoods, and deck lids. This rust first appears as a small bubble in the paint. If you push on this bubble with a sharp screwdriver, you may make a hole. Rust that starts on the outside is usually caused by a rock chip or scratch that exposes bare metal. In repairing a rust hole, you must cut out all of the soft metal and weld or bond in replacement metal. Outside rust can be repaired by sandblasting or grinding.

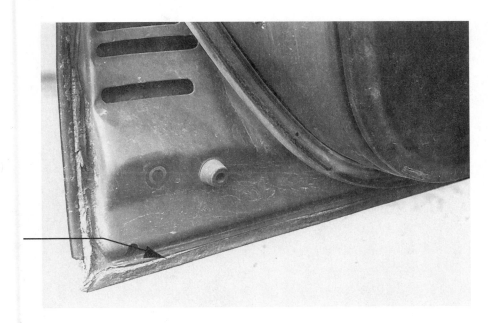

FIGURE 9–3   Hem flange rust on the inside of a door. This is often the first place a vehicle rusts. The flange separates from the door frame as rust forms between them.

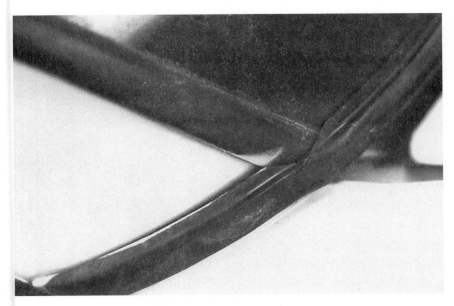

FIGURE 9–4   Overspray on the sail panel moulding of this Cavalier is evidence of a previous, poor-quality repaint.

A previous refinish may cause many problems if you paint over it. Some evidence of repainting are overspray on mouldings (Figure 9–4), paint edges on the inside surfaces such as door openings or hoods, sand scratch swellings, or bull's-eyes. Painting over a problem surface will magnify the problem. To avoid excessive paint build-up, you should consider removing all of the refinish paint.

Peeling is caused by a loss of adhesion between the topcoat and the primer. Usually peeling is found on upper surfaces: hood, roof, and deck lid (Figure 9–5). Peeling OEM paint can be scraped off with a razor blade. If the primer underneath has not peeled, the primer can be sanded, sealed, and repainted. If the primer is also peeling, strip all of the primer to the bare metal.

FIGURE 9–5 *The peeling OEM topcoat on this Lumina is caused by a loss of adhesion.*

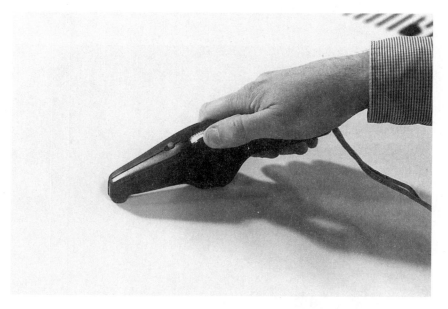

FIGURE 9–6 *A gauge is used to determine the thickness of OEM paint on a Dodge Dakota. In this example, the paint was 4.5 mils thick.*

After a close inspection of the paint, you can use a mil gauge or Tinsley gauge to determine the thickness of the paint (Figure 9–6). Most paint manufacturers recommend only one refinish on top of OEM paint. The total thickness (original plus repaint) should not exceed twelve mils. Too much paint causes those nasty problems like checking. If the mil gauge reads above 10, you should strip the surface to bare metal before refinishing.

9.2 — PLANNING

After a thorough evaluation of the surface, make a repair plan. Decide what areas are to be painted, blended, or melted, which mouldings need to be removed, and which areas are to be block-sanded. With

FIGURE 9–7 Quarter panel repair example.

FIGURE 9–8 Replacement fender example.

this plan in mind, the surface preparation will be more organized.

Here are two examples of plans. Figure 9–7 shows a quarter panel repair. The taillight and stripe will be removed. The quarter panel and rear bumper will be wiped with aggressive cleaner. The broken paint edge will be featheredged with 80-, 180-, and 320-grit paper. The bumper will be wet-sanded with 1,000-grit paper. The area above the stripe and in front of the filler will be back taped and masked for primer. The bare metal will be sprayed with self-etching primer.

Three coats of urethane surfacer will be applied. **Guide coat** will be dusted on the surfacer. The surfacer will be block-sanded with 320-grit dry paper followed with 600-grit wet paper. The sail panel will be sanded with 2,000-grit paper for a melt. The upper and front areas of the quarter panel will be sanded with 1,000-grit for a clear coat blend. After blowing off and wiping with a mild cleaner, the car will be final masked around the quarter panel with eighteen-inch masking paper and covered with a plastic sheet. Figure 9–8 shows a Taurus with the left fender replaced

FIGURE 9–9 *The mouldings, lock, and door handle have been removed from the door of this Mustang convertible. The repaint will be a high-quality tricoat finish.*

and the front bumper repaired. The plan for this vehicle is to remove the license plate bracket and to clean the hood, bumper, fender, and left front door with aggressive wax and grease remover. The fender will be sanded with 320-grit paper, and the bumper repair area will be sanded with 600-grit paper. The hood, door, and bumper will be prepped for a blend by sanding with 800-grit paper on a DA. The surfaces will be blown off and cleaned. The right fender, windshield, left front door window, and left rear door will be masked. The left fender will be masked off separately so it can be sprayed with sealer. In the booth, after the sealer is sprayed, the paper around the right fender will be removed, and the vehicle will be ready for painting and blending.

9.3 REMOVING OBSTRUCTIONS

A surface free of obstructions is easy to sand, paint, and buff. Unfortunately, a vehicle has many obstructions, including mouldings, bumpers, nameplates, taillights, and stripes. Obstructions should be removed to protect them from overspray and to provide access for painting. Some collision repair facilities have a policy that all mouldings, handles, locks, nameplates, and all

other obstructions are to be removed from the surface before painting (Figure 9–9). However, most shops do not get paid for the extra labor. These shops take off the easy-to-remove mouldings and nameplates and mask the difficult-to-remove mouldings. Other shops mask virtually all obstructions, removing only those that are impossible to mask (Figure 9–10). Some obstructions, such as the windshield reveal mouldings, cannot be removed. In these cases, there are ways to work around them. Remember that masking leaves a tape edge unless there is a gap between the panel and the obstruction.

Mouldings can be clipped, bolted, or glued on. The clipped mouldings are on rocker panels, doors, and occasionally on fenders and quarter panels. Look for screws on the inside of the door securing the ends. Push up on the bottom of the moulding and pull out at the top. The clip is designed to be compressed at the bottom and snapped into place at the top (Figure 9–11). Usually the clips are fastened to the body with welded studs. To remove the clip, tap it to one side with a hammer or screwdriver and push up to free it from the studs (Figure 9–12).

Bolted mouldings can be removed if you have access to the inside of the panel. On fenders, remove the fender liner. On doors, remove the trim panel. On quarter panels, remove the trunk trim. In most cases, to remove the door belt moulding, you will need to remove or loosen the window (Figure 9–13).

FIGURE 9–10   This pickup truck is masked before entering the spray booth. The windshield, running boards, and tires will be masked in the booth. Notice that all mouldings, handles, and other obstructions have been masked rather than removed in this quick repaint.

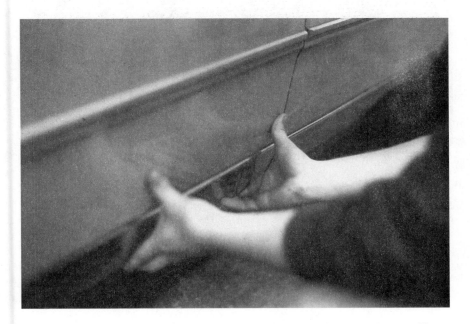

FIGURE 9–11   This clip-type moulding is removed by pushing up at the bottom, then pulling out on the top.

FIGURE 9–12 The clip is removed by pushing sideways.

FIGURE 9–13 The window was loosened to allow the removal of the door belt moulding from this Mustang.

Narrow glued-on rubber mouldings can be removed with a sharp putty knife. (Figure 9–14). Wide glued-on mouldings (two inches or more in width) can be removed with a heated putty knife. In some cases, the rubber moulding has a metal backing or metal insert. Be very careful not to bend the metal backed moulding during removal. Keep the moulding straight; otherwise it will be ruined.

Door handles and key locks can be removed from the inside of the door. For access, the door trim panel should be taken off. Bumpers that wrap around the fender or pickup bed should be removed.

FIGURE 9–14   A sharp putty knife is used to remove the mouldings from this F-150 pickup truck.

FIGURE 9–15   This S-10 pickup has minor collision damage to the left bedside, upper rear corner. In order to paint the bedside, the rear step bumper has been removed.

Otherwise, it is difficult to spray the area behind it (Figure 9–15). Chrome windshield mouldings can be removed with a tool that releases the clip, allowing the moulding to slide off (Figure 9–16). Pinstripes, made of tape can be easily removed with a razor blade. Larger tape stripes (one inch or more in width), should be warmed with a heat gun, then scraped off.

Some obstructions that are not removed are glued-in windshield mouldings and encapsulated glass mouldings. In many cases, nylon rope can be inserted between the body and the moulding (Figure 9–17). The gap that results between the moulding and the body will allow the paint to seep underneath the moulding, thereby eliminating a paint edge. The rope is removed after painting.

FIGURE 9–16 *Chrome windshield mouldings are removed by inserting a tool between the moulding and windshield and then releasing the retaining clip.*

FIGURE 9–17 *Nylon rope partially inserted to raise a glued-on windshield moulding. In actual use, the rope is completely inserted and not visible.*

9.4 CLEANING SURFACES

In order to prevent paint application problems, it is absolutely necessary to remove all contaminants from the surface to be repainted. The contaminants can be water soluble or oil soluble. The water-soluble conta-minants are removed with soap and water. Wash the car thoroughly. The oil-soluble contaminants are eliminated with a wax and grease remover. There are two types of wax and grease removers, aggressive and mild.

The **aggressive cleaner** is used to eliminate wax, road tar, gasoline, and other heavy contaminants. Use this cleaner before any sanding operations. If you sand with contaminants on the surface, you will drive

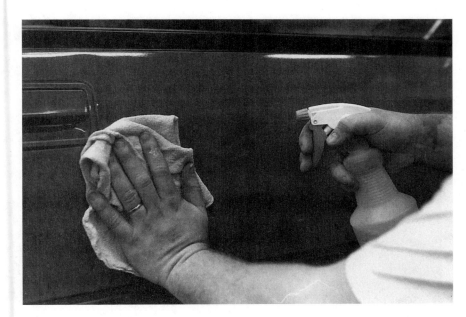

FIGURE 9–18 _Aggressive cleaner is sprayed on a contaminated surface. Then it is wiped off after a twenty-second reaction time._

them into the sand scratches, making them much more difficult to remove. To use an aggressive cleaner, load it into a spray bottle (Figure 9–18). Spray the surface to be cleaned, wait about twenty seconds, and then wipe it off with a clean towel. The cleaner must be wiped off when it is wet. Do not allow it dry on the surface. If it does dry, re-wet it and then wipe it off. To remove heavy road tar, soak a towel with the cleaner and hold it on the surface. This should soften the tar enough so it may be wiped off.

The **mild cleaner** is used to remove sanding sludge prior to painting. The mild cleaner evaporates rapidly, does not soak into the surface and cause sand scratch problems. It is applied in the same manner as the aggressive cleaner.

Most cleaners are solvent-based, meaning that they release VOCs when used. Water-based, and therefore low VOC, cleaners are available. There are also special cleaners for plastic parts.

9.5 — PREPPING

When the surface is thoroughly cleaned, you are ready for prep. There are several possible methods of prep:

- Strip to bare metal
- Strip to primer
- Sand

- Rust
- Weld
- Filler
- Aluminum
- Blend
- Melt
- Trim (new)
- Trim (used)

When the paint is in poor shape—for example, there is crazing, or there is too much paint—the paint should be stripped to the bare metal. When there is peeling of the topcoat, but the primer underneath is not peeling, the paint should be stripped to the primer.

There are four ways to strip to the bare metal:

- Grinding
- Sanding
- Chemical stripping
- Blasting

The grinding method uses a 24-grit disc on a buffer (Figure 9–19). Use this method if there is a repaint over OEM paint. This method removes paint, and if you are not careful, the metal. The grinder produces heat, enough to warp the panel if you are not careful. If you use this method on flat panels, always keep moving, work on one area briefly (one minute), then move to another area for one minute. Keep moving around to minimize the heat buildup. If the paint clogs the disc, slap the disc with a screwdriver blade to remove the buildup. Do not use this method over fiberglass or plastic.

FIGURE 9–19 A buffer fitted with a pad and a 24-grit disc is used to strip paint. This setup will aggressively remove paint or metal, so be careful.

FIGURE 9–20 A finish-sanding DA equipped with 80-grit paper is used to sand OEM paint.

The sanding method is used to strip OEM paint. A DA with 80-grit paper works well (Figure 9–20). Again, be careful of excessive heat buildup on flat panels. Also be careful on fiberglass and plastic edges.

Paint stripper is a chemical method of removing paint. Use this method if there is a lacquer finish that melts and smears instead of dry sanding. Check local rules about what to do with the waste created by chemical stripping. Mask vehicle seams to keep the stripper out of inaccessible areas. Wear protective gloves and work in a warm, well-ventilated area. Shake up the can of stripper prior to opening. Pour the stripper over the paint surface. Spread out with a paint brush. Move the stripper only once. Do not go back over the surface. Allow the stripper to loosen the

FIGURE 9–21 In this example of paint stripper, a small amount of stripper was applied to the surface and allowed to react for about ten minutes. The bubbled paint is removed with a putty knife.

FIGURE 9–22 A spot sandblaster is quite useful for removing rust.

paint. This may take several minutes. Scrape off the loose paint with a putty knife (Figure 9–21). Reapply the stripper to stubborn paint. Thoroughly rinse the stripped panel with water. Any trace of stripper will cause paint problems. Prime the bare metal immediately.

Blasting can be used to remove paint. Sandblasting uses abrasive sand at 100 psi to remove paint and rusty metal. This method can be used to strip vehi-cles, but it can easily warp, distort, and weaken metal. A small captive sandblaster (Figure 9–22) can be used to strip paint and to remove rust. This small blaster is also useful in cleaning a weld area before and after welding. Gentle blasting methods, plastic media blasting or soda blasting, use soft plastic granules at 30 psi. These methods are so precise that paint can be removed layer by layer. This service is usually offered by specialized shops.

FIGURE 9–23   A razor blade is used to strip the topcoat off the roof of this Lumina.

FIGURE 9–24   In this wet-sanding example, the sandpaper is wrapped around a sanding pad to insure uniform sanding pressure.

For peeling paint, use a razor blade to strip off the topcoat (Figure 9–23). Keep the razor blade at a low angle. Start at a peeled area and simply push the razor blade under the paint. Be careful not to dig the blade corners into the primer. If you do, it will leave a gouge that must be filled. Usually the hood, roof, deck lid and upper surfaces of the fenders, doors, and quarter panels are stripped in this manner. This process is remarkably fast, about one-half hour per panel. If there are not any gouges and the OEM primer is not peeling, you can sand the primer with 320-grit paper, seal, and paint.

If the paint surface is in good shape and not excessively thick, sanding will prepare it for a durable topcoat. As explained in Chapter 8, sanding can be done wet or dry. If you are wet sanding, always use a sanding pad to prevent finger troughs (Figure 9–24). Use plenty of water to lubricate the 400-grit paper. Check

FIGURE 9–25 _The left quarter panel on an Acura is probed to see if the metal is weak. It is._

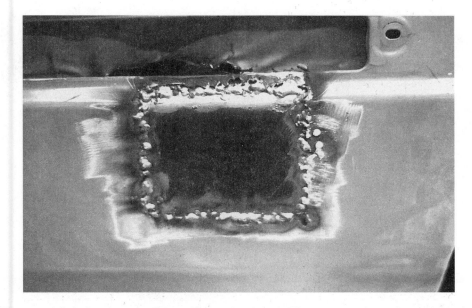

FIGURE 9–26 _The moulding has been removed from the Acura. The rust-weakened metal has been cut out and a patch panel MIG-welded into place._

the surface by using a squeegee to remove water. Sand until the shine is gone. For dry sanding, use 320-grit paper on a DA or jitterbug. Be careful around the edges, because it is easy to sand through them. Keep the sander flat to the surface.

If you encounter rust when you evaluate the surface, always probe with an awl or sharp screwdriver (Figure 9–25). Rust-weakened metal will often collapse, leaving a hole. Do not try to paint over rust bubbles; they will continue to rust. If the metal does not collapse, you can sandblast to remove all traces of black rust. Etch with acid to remove invisible rust and spray with self-etching primer or epoxy primer. If you need to fill gouges created by sandblasting, spray a coat of surfacer. After curing, apply two-part spot putty to the gouges, then sand and spray another coat of surfacer. If you have a hole from rust, cut out all of the soft metal and weld or bond in a patch panel. Be sure to prep the enclosed side of the repair (Figure 9–26).

FIGURE 9–27   An abrasive pad is used to remove paint. Unlike a grinder, the pad does not remove metal.

FIGURE 9–28   Plastic body filler and hardener are mixed together prior to an application on a dent.

Dents can be filled with plastic body filler. The filler should not be thicker than one-eighth inch. If the dent is deeper than one-eighth inch, continue metal finishing to raise the dent. Filler can be applied to bare metal or to cured epoxy primer. The procedures are similar. Metal-finish the dent to contour; shrink any stretched areas. Remove the paint with a grinder and a 24-grit disc (Figure 9–27). If the plan is to use epoxy primer, it is sprayed at this time. Allow the primer to cure; then scuff sand. Mix plastic filler on a board. Measure out the required amount of filler and hardener. Mix by stirring the filler and hardener together (Figure 9–28). Apply an overfill to the dent. After curing, air-file with 40-grit paper (Figure 9–29). In some cases, a two-part glazing putty is applied to the filler to prevent bleed-through of the hardener and to fill pinholes. The two-part glazing putty is mixed in the same way as the filler. After curing, sand with 80-grit

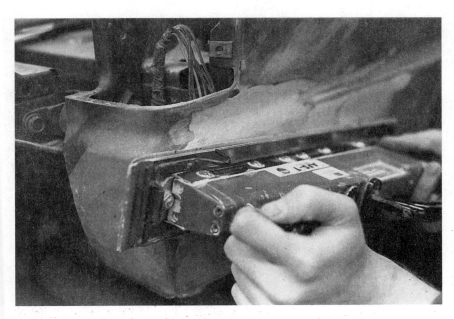

FIGURE 9–29 An air file with 40-grit sandpaper will quickly sand down the filler.

FIGURE 9–30 The fender of this Corsica has been repaired. Then the paint was featheredged with 80-, 180-, and 320-grit papers.

paper followed by 180-grit paper. Always check the filler surface for pinholes, gouges, or deep scratches. These should be filled before priming and surfacing. Featheredge the broken paint edge around the filler with 80-, 180-, and 320-grit papers on a DA, jitterbug, or block (Figure 9–30).

A weld is a corrosion hot spot requiring careful prep. If the weld bead is too tall, dress it down with a cutoff wheel. Stay on the weld; do not grind and weaken the surrounding metal. Finish dressing the weld with 80-grit paper on a DA. The heat of welding will blister the nearby paint. Remove the blistered paint with a nylon abrasive wheel (Figure 9–31). Featheredge the remaining paint. The back side of the weld must be prepped. Also use the nylon abrasive wheel to remove weld scale. If the back side of the weld is impossible to reach, blow it off as well as possible with compressed air. Spray the front and

FIGURE 9–31   This weld area is cleaned up with an abrasive wheel.

FIGURE 9–32   Sand fix paste and gray abrasive pad are used to abrade the clear coat on the hood of this Corsica. The hood will be blended.

back sides of the weld with two coats of epoxy primer. Surfacer may be sprayed to fill, if needed.

Aluminum prep is similar to steel prep. Epoxy primer is used over bare aluminum. Do not use a grinder to strip paint from an aluminum panel. The heat of the grinder will warp the sensitive aluminum. Use a DA and 80-grit paper to remove the paint. Apply epoxy primer over bare aluminum before using the plastic body filler.

To prep for a **blend,** one of three methods can be used. After the surface is thoroughly cleaned, the clear coat can be sanded to accept a refinish clear by sanding with 800-grit dry paper on a soft pad DA, wet-sanding with 1,000-grit paper, or scuffed with a gray abrasive pad and sand fix paste (Figure 9–32). Any one of these methods will abrade the clear coat to accept the refinish clear. The DA method is the fastest; the wet-sanding method is the messiest. Whatever

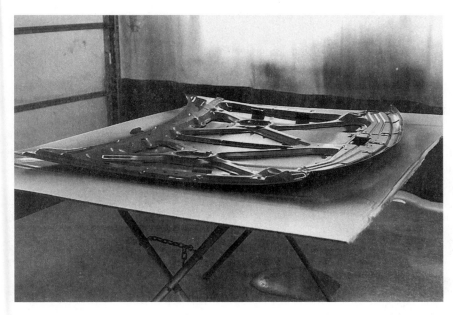

FIGURE 9–33 _This hood has been trimmed. The cardboard is used to prevent damage to the outside of the hood._

method is used, be sure that there aren't any shiny spots left. A shiny spot indicates an area that was not sanded. An unsanded area could lead to a loss of adhesion and paint failure. Check to see if the paint manufacturer recommends that an adhesion promoter be sprayed over the blend area.

The last type of surface prep is to trim metal replacement parts. Painted areas inaccessible from the outside should be painted before they are installed. These areas include door jambs, the underside of hoods, fender flanges, and undersides of deck lids. Replacement part sources can be OEM, aftermarket, or used.

To prep OEM and aftermarket parts, first unpack the parts and check for damage. Reject any dented parts. Wash and clean each part. Check each part to see if it has baked-on factory E-coat primer. While wearing gloves, saturate a rag with enamel reducer. Lightly   wipe the rag on the part. If, after several wipes, the primer softens and comes off on the rag, it is not E coat. A non–E-coat primer is not durable. Remove non–E-coat primer by wiping with reducer or sanding. Prime the bare metal with epoxy primer. If the part has E-coat primer, preserve as much of it as possible. Sand out runs and rough areas. On outer surfaces, such as the outside of a fender, sand with 320-grit paper on a DA. On inner surfaces, such as inside of a hood, sand with a red scuff pad. Make all of the shiny primer dull. Scuff whatever areas were painted on the original part, even if they will be covered and not readily visible. After the part is scuffed, blow off the dust and wipe with a mild cleaner. Take

the part to the booth and spray a coat of epoxy primer. After the proper flash time, spray a basecoat to hiding. Spray one coat of clear if needed (Figure 9–33).

Used parts or LKQ parts are commonly specified by insurance companies. When used parts are received, check them for damage, rust, excessive paint, or other paint defects. The trimming of used parts is similar to that of new parts in that the inaccessible areas must be painted. However, there will usually be some time required to remove mouldings, handles, locks, and decals. If the existing paint is in poor shape, it must be stripped. Sometimes the parts have inferior-quality repaired damage. In this case, it would be best to grind out the existing plastic filler, metal finish to within one-eighth of an inch of contour and refill.

9.6 — PRIMER MASKING

When the prep work is completed, the vehicle can be masked for primer. Masking for primer should be a quick and efficient task. Mask to prevent overspray damage but do not spend an excessive amount of time. For example, if you have prepped a door dent by filling, air filing, and featheredging, your primer and surfacer should cover all of the scratches. To keep the primer and surfacer where they belong, mask off the adjacent areas. If you lay down masking tape and

FIGURE 9–34   Notice the back-taped edge that is used to prevent the primer from entering the door opening.

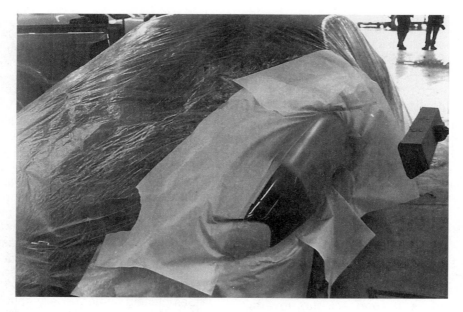

FIGURE 9–35   Plastic sheet used to mask for primer.

paper around the area to be primed and surfaced, you will end up with an edge of surfacer that is difficult to sand down. To avoid this problem, use the **back tape** method. Back tape means that you apply the tape, then roll up the edge nearest to the primer and surfacer area (Figure 9–34). When you spray, the rolled tape prevents a hard edge. This works best on edges and body lines. Always work from the inside

out when you mask. Tape and paper the inside of doors, hood, and deck lids. Then shut them and tape the outside. It is much easier and faster to tape and paper a car than it is to scrape or rub off overspray. After the immediate areas are masked, move the car to the spray booth. Place a plastic sheet or bag over the areas that will not be sprayed (Figure 9–35).

FIGURE 9–36 _Invisible rust is removed with metal conditioner. Use the abrasive pad to work the conditioner into the metal._

9.7 — PRIMING

For proper corrosion protection, the bare metal must be primed. The first step is to wipe the bare metal with aggressive cleaner. This will eliminate any oil that may have been deposited from the technician's bare hands onto the bare metal. Do not get the cleaner on the body filler. When the bare metal is cleaned, it is ready for one of the two priming methods. The more time-consuming way is to use a **metal conditioner** to clean the metal of microscopic rust, followed by a **conversion coating** to help the surfacer bind to the metal. The other method is to spray on a primer. There are two types of primers: self-etching primer and epoxy primer.

Follow these procedures for **metal conditioner** and conversion coat. Do not get any acid on the filler.

1. Cleaning with metal conditioner. Mix the appropriate cleaner with water in a plastic bucket according to label instructions. Usually the ratio is one part metal conditioner and two parts water. If rust is present, work the surface with a stiff brush or abrasive pad (Figure 9–36). Keep the metal wet for the specified amount of time. Do not allow the surface to dry. After the specified amount of time, rinse the bare metal with water or wipe dry, depending on product instructions.

2. Applying conversion coatings. Pour the appropriate conversion coating into a plastic bucket. Using an abrasive pad, brush, or spray bottle, apply the coating to the metal surface (Figure 9–37). Leave the conditioner on the surface two to five minutes. Apply only to an area that can be coated and rinsed before the solution dries. If the surface dries before the rinsing, reapply. Flush the coating from the surface with cold water. Wipe dry with a clean cloth and allow to air-dry completely. The desired surfacer can now be applied.

On bare metal areas that have not been exposed for more than a few hours, a self-etching primer or epoxy primer can be used. These primers will bond with the bare metal and provide a good base for the surfacer.

Self-etching primer usually consists of base that is reduced with phosphoric acid (Figure 9–38). Follow the paint manufacturer's recommendations on mixing. Spray on a wet coat over all bare metal and filler. Allow the first coat to flash and spray on another wet coat. Clean the spray gun immediately after use. The mixed self-etching primer is most effective if used within eight hours of mixing.

The other spray method of priming is to use an epoxy primer (Figure 9–39). This method is recommended for use on steel, galvanized steel, and aluminum. Read the manufacturer's directions. Usually there are two or three components to be mixed. Some epoxy primers have an induction time. This is the

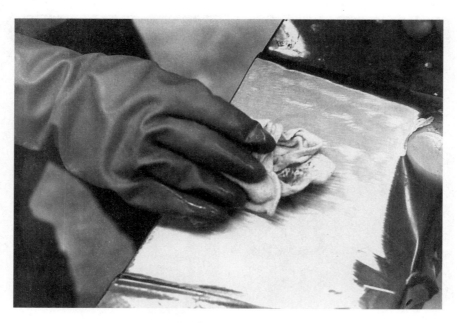

FIGURE 9–37 Conversion coating will allow the surfacer to properly bind. In this case, the surface is kept wet with conversion coating for a specified time.

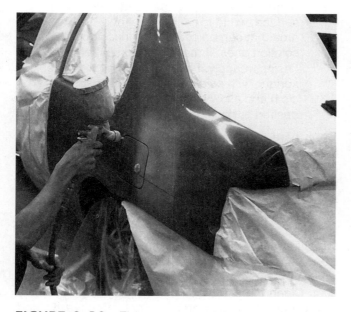

FIGURE 9–38 This two-part self-etching primer is mixed one-to-one and two wet coats are sprayed on the bare metal.

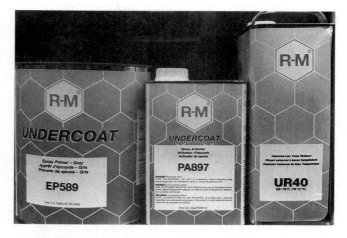

FIGURE 9–39 This epoxy primer is mixed 4:1:1. It is sprayed on in the same manner as self-etching primer, except it does not need to be sprayed as wet as self-etching primer.

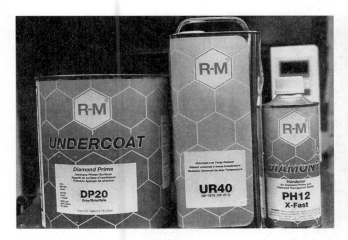

FIGURE 9–40 *This urethane surfacer is mixed 4:1:1. It offers high build and limited shrinkage.*

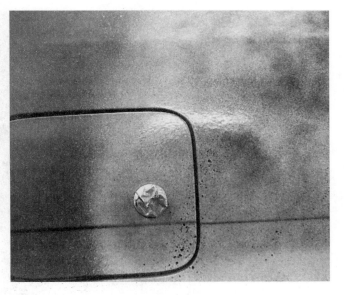

FIGURE 9–41 *The dark speckles over the surfacer are guide coat. The guide coat and surfacer are block-sanded. Guide coat remaining after block sanding indicates low spots.*

amount of time required for the mixed components to react before they can be sprayed. Mix the components and allow the induction time, if necessary. Then spray on a medium wet coat. After flash time, spray on another medium wet coat. Clean the paint gun immediately.

9.8 — SURFACING

After the metal has been primed, a **surfacer** should be used to fill in all scratches and depressions (Figure 9–40). A good rule of thumb is to spray on three coats of urethane surfacer. Usually the surfacer has three components. Follow the manufacturer's recommendations when mixing. Spray on three coats, observing the proper flash time between coats. Clean the paint gun out immediately after use.

In order to make topcoat coverage faster, some surfacers can be tinted. If the surfacer is similar in color to the topcoat, fewer coats of color are needed.

After the last coat of surfacer has flashed, dust on a contrasting color guide coat (Figure 9–41). As this guide coat is sanded, low areas and scratches will be identified. Do not skip the guide coat step. The urethane primer may need to cure for as long as eight hours in some cases. Production shops can use an infrared drying system to speed up the cure time.

9.9 — BLOCK SANDING

After the surfacer has cured, the repair can be **block-sanded.** If this block sanding is done correctly, the repair will be invisible under the topcoat. If done incorrectly, there will be an obvious ring around the repair. Start with 320-grit dry paper on an inflexible holder. A flat wooden paint stick or wooden ruler works well. Hold the stick as shown in Figure 9–42. Sand until all of the guide coat is gone. Bare metal indicates a high spot. Use a body hammer to lower the area or re–metal finish. If guide coat remains after you have sanded all the guide coat surrounding it, you have located a low spot. Span the area with the paint stick. If the surfacer has filled the low spot, you can carefully block-sand and remove the guide coat. If bare metal appears around the low spot as you sand, the low spot cannot be blocked out. It must be filled with body filler.

If scratches remain in the surfacer after block sanding, you can fill them with two-part spot putty. Use a razor blade to apply the mixed putty. The razor blade minimizes the amount of putty applied, making the

FIGURE 9–42   A flat panel stick is wrapped with 320-grit paper. This method of block sanding works quite well. Turn the stick on edge to block-sand high-crown surfaces.

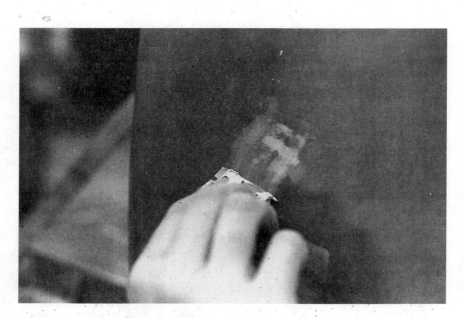

FIGURE 9–43   Use a razor blade to apply two-part spot putty.

sanding faster (Figure 9–43). When you sand the putty, be sure not to sand down the adjacent surfacer. Start with 80-grit paper, then 180-grit paper. Apply two coats of surfacer.

After the surfacer has been block-sanded flat, you can finish sanding with 600-grit wet paper. Because the surface is now flat, you can use a hand pad or sanding block. Remove all of the 320-grit scratches. Sand out beyond the surfacer edge to provide a tapered surface.

9.10 — FINAL CLEANING

This final cleaning of the car must be thorough. Remove all of the masking materials. Wash the car to remove all of the dust and sanding residue. Use compressed air to blow out all doorjambs, cracks, mouldings, or anywhere that water and sanding

FIGURE 9–44 *Commonly used masking tapes: two-inch, three-quarter-inch, one-quarter-inch, and one-quarter-inch fine line.*

sludge may be hiding. On one-panel jobs, you can probably just carefully blow the car off. Finally, use a **mild cleaner** to wipe down all surfaces to be painted or clear coated. The vehicle is now ready for paint masking.

9.11 — PAINT MASKING

Proper masking will make the repair invisible. Overspray on mufflers, doorjambs, wheel wells, frames, and mouldings is the sign of a low-quality repair. Masking is a very important step in the painting preparation process. Masking keeps paint overspray from contacting areas other than those that are to be repainted. This has become even more important since the popular use of acrylic urethane and two-component–type paints. Once these types of paints dry, the over spray cannot be removed with a thinner or other solvent. These paints will have to be removed with a compound or other time-consuming means.

The basic materials for any masking job are masking paper and tape. Automotive paper comes in various widths from three to thirty-six inches. Automotive masking paper is heat resistant so that it can be used safely in baking ovens. It also has good wet strength, freedom from loose fibers, and resistance to solvent penetration. *Never* use newspaper for masking a ve-

hicle, since it does not meet any of these requirements. Newspaper also has the added disadvantage of containing printing inks that are soluble in some paint solvents. These can be transferred to the underlying finish, causing staining.

Automotive masking tape comes in various widths from one-quarter to two inches. The most frequently used tapes are shown in Figure 9–44. Larger-width tapes are used only occasionally, since they are expensive and difficult to handle. Automotive masking tape should not be confused with tapes bought in the hardware or paint stores for home use. The latter will not hold up to the demanding requirements of automotive refinishing. It is interesting to note that the average-size vehicle takes two to two and one-half rolls of tape to be completely masked.

Plastic masking tape is available in several widths from one-eighth to one-half inch. It easily conforms to any shape. This type of tape is most often used as an edge tape on mouldings and windshield gaskets.

The use of masking paper and tape dispensing equipment (Figure 9–45) makes it easy to pull and tear the exact amount of paper needed. Some masking machines permit tape to adhere to one or both edges of paper as it is rolled out.

There are several types of masking covers available. One type of cover is the plastic tire cover that eliminates the need for masking off the tire. Others include a body cover and liquid mask. Liquid mask is a sprayable material that dries and protects areas from overspray. It also holds down the dirt. After the car is

FIGURE 9–45 *This masking tape machine is set up with eighteen-inch and twelve-inch paper.*

painted, the dried liquid mask is removed by washing the car. Before any masking materials are applied, the vehicle must be completely cleaned and all dust blown from the vehicle. The masking tape will not stick to surfaces that are not clean or dry. It is most important that the tape be pressed down firmly and adhered to the surface. Otherwise, paint solvents will creep under the tape. In the case of a two-color job, where the color break is not hidden by a capping strip or moulding, it is *vital* that the masking tape is firmly pressed down.

SHOP TALK

It is wise to completely detail (that is, check to see that all prior steps have been performed) the car before the masking job is started, and, of course, after the paint job is completed. Reason: Improper masking can also cause a dirty paint job if done over a dirty car.

Foam tape can be used to quickly mask off the insides of doors, hoods, and deck lids. Simply pull out and cut the required amount. Open the panel and apply the foam tape to the edge. The foam tape prevents overspray from entering and leaves a soft paint edge (Figure 9–46).

If the paint shop is cold and damp with little air movement, the masking tape probably will not stick to glass or chromium parts because of an almost invisi-

ble film of condensation that has formed on these parts. It must be wiped off before the tape will adhere properly.

Masking tape generally will not stick on the black rubber weatherstripping used around the doorjambs and deck lid opening. To mask rubber weatherstripping, apply clear lacquer thinner with a rag, allow it to dry completely, and then apply the tape. When masking doorjambs, be sure to cover both the door lock assembly and the striker bolt.

Although masking tape has an elastic property, it should be stretched only on curved surfaces. This is especially true when masking newly applied finishes that are still soft beneath the surface. Another reason to avoid stretching the tape is that it can increase the degree of tape marking on the finish.

Most experienced maskers find it easier to hold and peel the tape in one hand while they use the other hand to guide and secure it (Figure 9–47). This gives tight edges for good adherence and allows the masker to change directions and go around corners with tape. To cut the masking tape easily, quickly tear it against a finger (Figure 9–48). This procedure will permit a clean cut of the tape without any stretching.

Be careful that the tape does not overlap any of the areas to be painted. Loop or overlap the inner tape edge to make, and follow, curves. The tape will stretch to conform to curves. Difficult areas such as wheels can be masked using this process, but more often wheel covers are used to save time.

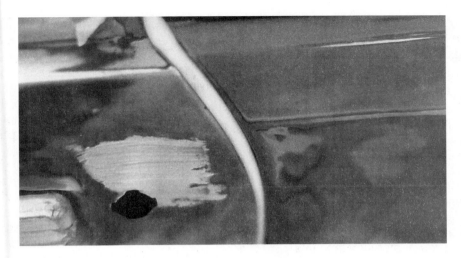

FIGURE 9–46   Foam tape is used to seal the door opening to prevent primer entry.

FIGURE 9–47   For fast tape application, hold one end in place and pull the roll.

FIGURE 9–48   Tear masking tape by holding in place with a finger.

Here are some general recommendations on masking paper and tape size for the various areas of a car to be masked:

- Mask antennas by making a sleeve with pre-taped three-inch masking paper and secure at the base with masking tape. On windshields, use two widths of either fifteen- or eighteen-inch masking paper. The top layer must overlap the bottom to prevent overspray.
- Rear windows are masked similarly to windshields with the use of two widths of fifteen- and eighteen-inch masking paper.
- Apply twelve- or eighteen-inch masking paper to windows for fast, economical protection. On door handles, apply three-quater-inch tape in a lengthwise continuous strip to insure faster removal. Chrome drip rails and mouldings also require three-quarter-inch or wider widths of masking tape. Outside mirrors can be masked with two-inch tape or six-inch masking paper.
- The wide variety of shapes and widths of grilles and bumpers might require various widths of masking paper; three, six, twelve, and eighteen inches are the most popular widths to use. Mask protective side mouldings and wheel well mouldings with two strips of three-quarter inch or two-inch as required. For protecting the tire and wheel, wrap adjoining pieces of eighteen-inch masking paper around the tire, Taillights are masked with six- or twelve-inch masking paper. Mask letters and emblems with one-eighth-inch or one-quarter-inch tapes. A pocket knife is a handy tool to work the tape into place on such small items. If these items are too difficult to mask, they can often be removed during the masking operation and replaced after refinishing. Use two or three pieces of thirty-six-inch masking paper to protect the inside of the trunk area. Use lacquer thinner to promote tape adhesion when masking weatherstrip. Mask around doorjambs with 6-inch pretaped masking paper.

When one is masking large areas such as bumpers, it is easier to manage the paper if it is tacked in the middle of the bumper with tape first. Then each side can be masked without the paper dragging on the floor and getting in the way.

Before masking glass areas, remove such items as wiper blades. The wiper shafts can be protected in the same manner as radio antennae and door handles. Glass areas themselves can be masked by first applying tape along the very top and edges of window mouldings. Then use two pieces of masking paper to cover the glass area. The tape on the edge of the paper should overlap the tape placed on the moulding. The top piece of paper should overlap the bottom layer of paper. If necessary, fold and tape any pleats in the paper so that there is no dust seepage. To mask lights that cannot be completely covered by tape or light covers, masking paper can be cut, folded, and worked around before being held down with tape.

One-sixteenth and three-thirty-seconds-inch fine line tape can be used to protect existing stripes from overspray or damage to adjacent panels. Use fine line tape for precise color separation in two-tone painting and for painting vivid, clean stripes. Its added flexibility and conformability makes the painting of curved lines easier, with less reworking.

Inspect the masking very carefully for any over-masked or under-masked areas that will make extra work after the vehicle is painted. Over-masked areas mean that the painter must touch up the part of the car that should have been painted. Under-masked areas must be cleaned with a solvent to remove overspray that detracts from the overall appearance of an otherwise good job. Figures 9–49 and 9–50 show examples of vehicles masked for paint.

FIGURE 9–49 *This Lumina is masked for the painting of the roof and right fender. The right front door will be blended. The front bumper has been removed for separate painting.*

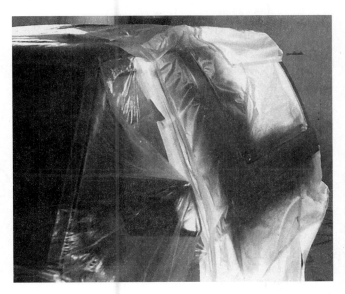
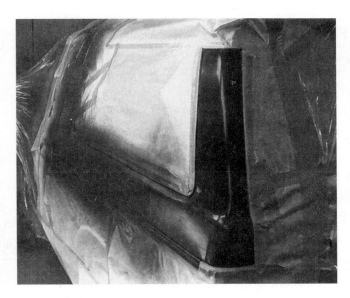

FIGURE 9–50 The entire left side of this Blazer was refinished. During assembly, the refinish paint was scratched. This is the masking that was used in the repair of the scratch. The body lines were back-taped.

9.12 — SEALING

The last step in surface prep is sealing. A **sealer** is used as an adhesion promoter to make a strong, durable bond for the topcoat. The sealer is often an epoxy primer. When the epoxy primer is used as a sealer, the mixing ratio may be different, so check the refinish manual for instructions. After trimmed replacement parts are installed, the outside surfaces are sprayed with a sealer and then with topcoat (Figure 9–51). On a complete repaint, a sealer is sprayed over the sanded OEM paint to make a good bond for the topcoat. In these two operations the sealer is a wet on wet application, no sanding required. Be sure to allow the required amount of flash time between sealer and topcoat. Some shops spray a sealer over sanded urethane surfacer prior to topcoat. In this case a sealer is optional.

FIGURE 9–51 Sealer is sprayed on this Cadillac replacement fender prior to topcoat.

REVIEW QUESTIONS

1. Technician A says that aggressive wax and grease remover is used to remove road tar. Technician B says that a mild wax and greaser remover is used to remove sanding sludge. Who is right?
   a. technician A
   b. technician B
   c. both technician A and technician B
   d. neither technician A nor technician B
2. Metal conditioner is used to remove paint.
   a. True
   b. False
3. Technician A sands for a blend with 800-grit paper. Technician B uses 400-grit paper to sand for a blend. Who is right?
   a. technician A
   b. technician B

c. both technician A and technician B
d. neither technician A nor technician B
4. A 24-grit disc is used to remove
   a. paint.
   b. rust.
   c. filler.
   d. a & b only
5. Technician A sandblasts rust. Technician B grinds rust. Who is right?
   a. technician A
   b. technician B
   c. both technician A and technician B
   d. neither technician A nor technician B
6. Usually _____ coats of surfacer are sprayed on filler.
7. Technician A removes all obstructions before a repaint. Technician B masks all obstructions except those impossible to mask. Who is right?
   a. technician A
   b. technician B

c. both technician A and technician B
d. neither technician A nor technician B
8. Which of the following is used to sand clear coat for a blend?
   a. 800-grit dry paper
   b. 1,000-grit wet paper
   c. Gray abrasive pad
   d. All of the above
9. Technician A preps for a melt with 1,000-grit paper. Technician B uses 2,000-grit paper to prep for a melt. Who is right?
   a. technician A
   b. technician B
   c. both technician A and technician B
   d. neither technician A nor technician B
10. Dried liquid mask is removed with water.
   a. True
   b. False

Chapter

10

Paint Mixing Systems

Objectives

After reading this chapter, you should be able to:
- Describe the OEM color selection process.
- Identify paint mixing system components.
- List steps in mixing a color.

Key Term List

tint
binder

 **OEM COLOR SELECTION PROCESS**

The color selection process begins with various paint manufacturers examining trends in color preferences. Using these preferred colors, they formulate paint. The new colors are sprayed onto test panels. The vehicle manufacturers decide which colors they want to have on the next year's vehicles from these test panels. After the next year's colors are chosen, each vehicle manufacturer allows the paint manufacturer's to submit bids on all colors, regardless of which paint manufacturer submitted the test panel. The successful bidder provides the color to the vehicle manufacturer.

 **PAINT CODES**

All vehicles have a paint code. This code can be found in various locations, depending on vehicle type (Figure 10-1). The code will specify the vehicle color. If the vehicle is a two-tone, both colors will be listed. Often a trim code for a secondary color, such as bumper fascia, will be listed. The code may also specify what type of paint the vehicle has (Figure 10-2).

The paint code is used to determine the paint manufacturer's part number. Each paint manufacturer publishes a color book. This book is divided by vehicle make, and each make is divided into years (Figure 10-3). Simply find the appropriate vehicle make and year page. Then look up the paint code. This will list the paint part number (Figure 10-4).

Using the paint part number, you can access the paint formula. One method of access has the paint formulas on microfiche (Figure 10-5). To find a formula, select the appropriate microfiche, and load it

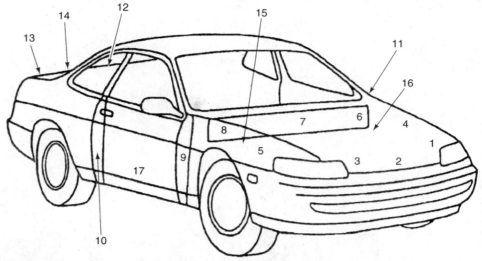

Model	Position	Model	Position
Acura	9	Hyundai	6,7
Alfa Romeo	4,13	Isuzu	2,10
AMC	9,10	Lexus	7,8
Audi	12,13	Mazda	1,2,3,4,6,8
Austin Rover	17	Mercedes	2,7,9
BMW	4,5	Mitsubishi	7
Chrysler	3,5,16	Montero / Pickup	3
Chrysler Corp	3,5	Cordia / Tredia	4
Caravan / Voyager / Ram Van	6	Others	1,2,3
Chrysler Imports	1,2,4	Nissan	1,3,4,6,8,15*
Colt Vista	16	Peugeot	2,3,4,5,8
Conquest	7	Porsche	9
Diahatsu	1,6,7	Renault	1,3,4,5,8
Datsun	2	Rover	1,3,4,5
Dodge D50	3	Saab	5,6,8
Ford	10	Subaru	2
Ford Motor Co	10	Suzuki	7,11
General Motors		Toyota Passenger	7,8,14
A, J and L Bodies	14	Truck	4
E and K Bodies	12	Volkswagen	2,11
B,C,H and N Bodies	13	Volvo	6,7,8
GM Imports	2,12,13,14	Yugo	12
Honda	8,10		

FIGURE 10–1   *Paint code location.*

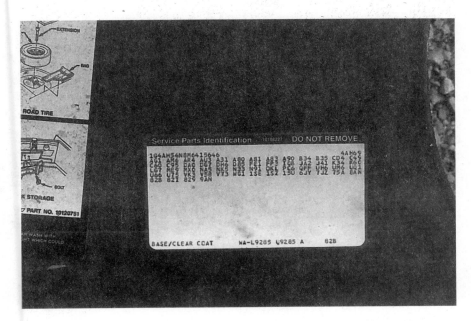

FIGURE 10–2 This Buick has a basecoat/clear coat finish. The paint code is 15U or WA-L9285.

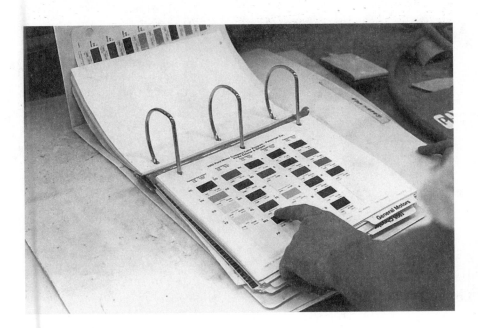

FIGURE 10–3 Paint color book.

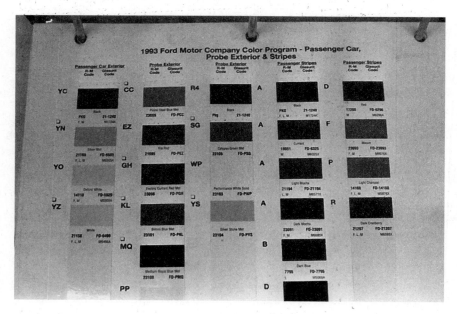

FIGURE 10–4 *The paint code is KL, and the paint part number is D23101.*

FIGURE 10–5 *The appropriate microfiche is selected.*

into the microfiche reader (Figure 10-6). Find the proper paint part number on the microfiche; the formula will be listed (Figure 10-7). You may notice that there are several alternates listed for the same paint part number. This is because the paint manufacturer has recognized a variance in color between vehicles with the same code. Alternate formulas are derived to match the different colors.

Paint formulas can also be accessed with a computer (Figure 10-8). The formulas are stored on a disk. The paint code is entered and the formula is listed on the screen. The formula can be printed out (Figure 10-9). There are formulas for pint, quart, two quarts, three quarts, and gallon amounts. If you need an amount of paint to test a color, mix one-tenth of a quart. Just move the decimal point one place to the

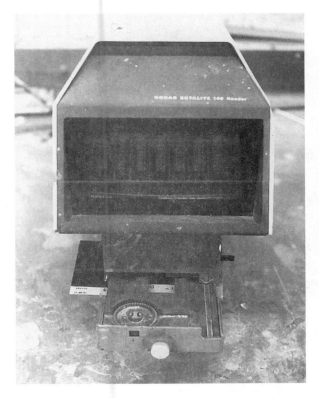

FIGURE 10–6 The microfiche is loaded into the reader.

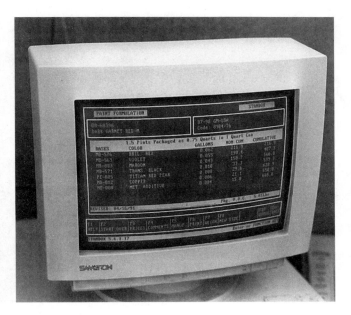

FIGURE 10–8 Paint formula on the computer screen.

D23101

1993 FORD AMERICA
BIMINI BLUE MET

Code: KL
OEM: Basecoat/Clearcoat

BASE	VOLUME	1 PINT	1 QUART	2 QUARTS	3 QUARTS	GALLON
UR50	100	41.5	83.0	165.9	248.9	331.9
BC151	530	228.1	456.2	916.4	1374.6	1832.9
BC405	795	345.0	690.1	1380.1	2070.2	2750.3
BC140	891	386.5	772.9	1545.9	2318.8	3091.7
BC300	932	404.2	808.5	1617.0	2425.5	3234.0
BC190	949	413.1	826.2	1652.5	2478.7	3305.0
BC200	969	421.8	843.6	1687.0	2530.6	3374.1
BC505	978	425.7	851.4	1702.9	2554.3	3405.8
BC101	1000	435.6	871.2	1742.3	2613.5	3484.7

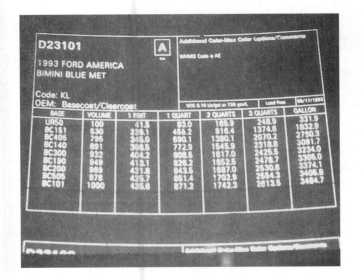

FIGURE 10–7 The formula for D23101. Amounts for one pint, one quart, two quarts, three quarts, and one gallon are listed.

AUTOMOTIVE FINISHES
MM240K ALT:1 – CHROMABASE
MIX SIZE: 2 QUART Ref. Cost: 72.10
3:05PM

806J	HS BLACK	164.8
815J	MULTIGRAD ALUM	246.8
862J	TRANSP RED	296.0
829J	LIGHT BLUE	338.0
813J	MED COARSE AL	377.2
802J	LS WHITE	404.0
816J	MED FINE ALUM	422.0
150K	B/C BALANCER	1383.2
175K	BINDER	1736.4

05/12/98 MIXED BY: _________ ** MSDS LEAD-FREE **

CUST: SMASH PALACE

FIGURE 10–9 Paint formula printout.

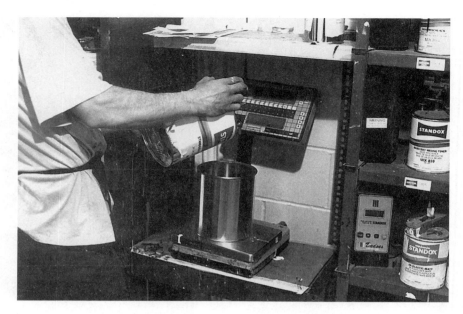

FIGURE 10–10   *Technician adding tint on an electronic scale.*

FIGURE 10–11   *Set scale to zero by pushing the button on top.*

left. An electronic scale is used to weigh out the specified amount of **tint** (Figure 10-10). The scale can be calibrated by weighing a nickel. The nickel should weigh 5 grams. The appropriately sized container is placed on the scale. The scale is set to zero before adding tint by pushing the button on the top (Figure 10-11). As tint is added to the can, the scale weighs the amount. Stop when the required amount has been added.

10.3 — TINTS

Tints are combined in specified amounts to create colors. Some tints are concentrated or high strength (Figure 10-12). A small amount can change the color. Low-strength tints do not change the color as much

FIGURE 10–12 High-strength green-blue tint.

FIGURE 10–14 Metallic tint.

FIGURE 10–13 Low-strength black toner.

FIGURE 10–15 Pearls can be liquid or powder.

as do high-strength tints (Figure 10-13). Often a tint will have a primary color and an underhue. Iridescent tints contain aluminum flakes (Figure 10-14). These flakes can be small (fine) or large (coarse) or sizes in between. Pearls are made from mica (Figure 10-15).

Frequently used tints are supplied in gallon cans in the mixing bank. Less commonly used tints are supplied in quarts. The mixing bank is a rack to hold the tint cans and mechanical stirrers to mix the tints (Figure 10-16). The mechanical stirrers should be run

FIGURE 10–16   Mixing bank.

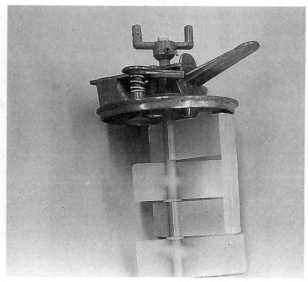

A

each morning and afternoon for fifteen minutes. Before a color is mixed, the tints should be stirred for five minutes. The stir lid takes the place of a paint can lid. The stir lid consists of a spout, closure, paddle, and paddle shaft (Figure 10-17). The paddle shaft engages the mechanical stirrer on the mixing bank. The can snaps in place using clips on the mixing bank shelf.

10.4 — MIXING

Mixing a color consists of the following steps:

1. Determine the paint code from vehicle information.
2. Determine the paint part number from the paint code.
3. Determine the paint formula from the paint part number.
4. Stir tints.
5. Obtain the appropriately sized container.
6. Place the container on the scale. Zero the scale.
7. Add the specified amounts of tints to make color. Figure 10-18 shows a color formula. The first tint, 818J, is added to 39.2 grams. The next tint, 827J, is added to 73.7 grams. Each tint is cumulative.

B

FIGURE 10–17   (A) Stir lid removed from the can. (B) Stir lid in use.

```
             UFTRING CHEVROLET
   B8868F H ALT:2 - CHROMAPREMIER BC
       MIX SIZE: SPECIAL (PINT (-)) Ref. Cost: 22.39
              5:07PM
   818J    BRIGHT ADJUST         39.2

   827J    BLUE                  73.7

   806J    HS BLACK              86.5

   820J    VIOLET                94.2

   813J    MED COARSE AL        100.0

   828J    HS FAST BLUE         104.8

   62320F  BINDER               233.9

   62330F  BALANCER             335.7


   05/12/98   MIXED BY: RJ    ** MSDS LEAD-FREE **
```

FIGURE 10-18 Paint formula listing cumulative weight of various tints.

To avoid problems:

- Clean stir lids before pouring. Remove the dried paint.
- Add tints to the center of the can, so it stays in a mass. If an overpour occurs, quickly remove some of the overpoured tint with a knife blade.
- Add tints slowly to avoid overpour.
- Put the mixed paint on a shaker for five minutes.

Some computers can compensate for an overpour. The overpour amount is entered and the computer re-calculates the formula. Additional tints are then added to correct for the overpour.

SHOP TALK

Not all collision repair facilities own a paint mixing system. Those facilities without mixing systems usually obtain paint from a jobber. The jobber will have a mixing system to formulate paint. This type of paint is called store mix. Another type of paint the jobber may have available is known as factory pack. Factory pack paint is mixed by the paint manufacturer and shipped as a ready-to-reduce product.

REVIEW QUESTIONS

1. Technician A says that a vehicle paint code may be found inside the driver's door. Technician B says that the paint code may be found in the trunk. Who is right?
   a. technician A
   b. technician B
   c. both technician A and technician B
   d. neither technician A nor technician B
2. The vehicle paint code may be found.
   a. on the radiator support.
   b. on the cowl.
   c. in the glove compartment.
   d. all of the above.
3. Technician A says that paint formulas are stored on microfiche. Technician B says that paint formulas are stored on a computer. Who is right?
   a. technician A
   b. technician B
   c. both technician A and technician B
   d. neither technician A nor technician B
4. Tints should be stirred at least
   a. twice a day.
   b. Twice a week.
   c. Twice a month.
5. Technician A calibrates the electronic scale with a dime. Technician B calibrates the electronic scale with a nickel. Who is right?
   a. technician A
   b. technician B
   c. both technician A and technician B
   d. neither technician A nor technician B
6. _____ are made from mica.
7. Technician A says that aluminum flakes can be various sizes. Technician B says that fine aluminum flakes are the largest aluminum flakes. Who is right?
   a. technician A
   b. technician B
   c. both technician A and technician B
   d. neither technician A nor technician B
8. Overpour means that too much of a tint was added when mixing paint.
   a. True
   b. False
9. Technician A adds tints to the center of the can. Technician B stirs the tints before use. Who is right?
   a. technician A
   b. technician B
   c. both technician A and technician B
   d. neither technician A nor technician B
10. Paint formulas are cumulative.
    a. True
    b. False

Chapter

11

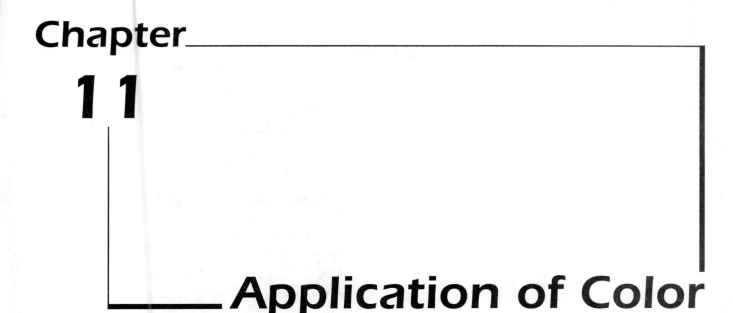

Application of Color

Objectives

After reading this chapter, you will be able to:

- **Explain how to mix paint.**
- **Identify the three types of paint: single-stage, basecoat/clear coat, and tricoat.**
- **Identify the three painting methods: spot, panel, and overall.**
- **Describe how to make a blend.**
- **Describe how to make a melt.**

Key Term List

front blend
back blend
flash time
melt

11.1 — PAINT MIXING

Paints are designed to be sprayed at a specified viscosity. Viscosity is the consistency of paint. Topcoat paint materials are usually shipped at as high a viscosity as practical, to minimize settling. In order to apply these materials, one must delute them to a sprayable viscosity. Incorrect viscosity will result in many paint problems. The specified viscosity is obtained by adding the proper amounts of the paint components. Figure 11–1 shows a paint label from a two-component paint; Figure 11–2 shows a paint label from a three-component paint. The amounts may be listed as ratios (4:2:1) as percentages (50% reduction), or as combinations (4:2:30%).

A ratio mixture is obtained by adding the specified amount of each part. The amount of each part could be one-half pint, one pint, one quart, or any other amount, as long as the parts are equal. For example,

FIGURE 11–1 *The boxes on this paint label contain mixing and application information. The upper box lists the reduction: 1 part paint to 1½ to 2 parts reducer. The next box indicates 2 to 3 coats are needed. The third box gives the flash time between coats: 20 minutes. The last box refers the user to product information form P-152. This form gives complete directions for successful application. (Courtesy of PPG.)*

211

BASE COLOR

CONTAINS: amyl acetate, 628-63-7; butyl acetate, 123-86-4; xylene, 1330-20-7; acrylic polymer, 96591-17-2. When mixed with activator also contains aliphatic polyisocyanate and hexamethylene diisocyanate monomer, 822-06-0.

WARNING: This product contains chemicals known to the State of California to cause cancer and birth defects or other reproductive harm.

(For further information refer to Material Safety Data Sheet)

WARNING: Vapors and spray mist harmful if inhaled. May cause eye and skin irritation. Can be absorbed through the skin. Flammable liquid.

IMPORTANT: May be mixed with other components. Mixture will have hazards of both components. Before opening the packages, read all warning labels. Follow all precautions.

ATTENTION: Cancer hazard. Contains material that can cause cancer.

NOTICE: Repeated and prolonged overexposure to solvents may lead to permanent brain and nervous system damage. Eye watering, headaches, nausea, dizziness and loss of coordination are signs that solvent levels are too high. Intentional misuse by deliberately concentrating and inhaling the contents may be harmful or fatal.

Do not breathe vapor or spray mist. Do not get in eyes or on skin.

WEAR A POSITIVE-PRESSURE, SUPPLIED-AIR RESPIRATOR (NIOSH/MSHA TC-19C), EYE PROTECTION, GLOVES AND PROTECTIVE CLOTHING WHILE MIXING ACTIVATOR WITH ENAMEL, DURING APPLICATION AND UNTIL ALL VAPORS AND SPRAY MIST ARE EXHAUSTED. Follow respirator manufacturer's directions for respirator use.

INDIVIDUALS WITH HISTORY OF LUNG OR BREATHING PROBLEMS OR PRIOR REACTION TO ISOCYANATES SHOULD NOT USE OR BE EXPOSED TO THIS PRODUCT. Do not permit anyone without protection in the painting area.

Keep away from heat, sparks and flame. **VAPORS MAY IGNITE EXPLOSIVELY.** Vapor may spread long distances. Prevent build-up of vapor. Extinguish all pilot lights and turn off heaters, non-explosion proof electrical equipment and other

sources of ignition during and after use and until all vapor is gone. Do not transfer contents to bottles or other unlabeled containers. Close container after each use. Use only with adequate ventilation.

FIRST AID: If affected by inhalation of vapor or spray mist, remove to fresh air. In case of eye contact, flush immediately with plenty of water for at least 15 minutes and call a physician; for skin, wash thoroughly with soap and water. If swallowed, **CALL A PHYSICIAN IMMEDIATELY. DO NOT** induce vomiting.

IN CASE OF: FIRE – Use water spray, foam, dry chemical or CO_2.

SPILL/WASTE – Absorb spill and dispose of waste or excess material according to Federal, State and local regulations.

REFER TO THE MATERIAL SAFETY DATA SHEET

KEEP OUT OF REACH OF CHILDREN   PHOTOCHEMICALLY REACTIVE

DIRECTIONS FOR USE

MIXING: ChromaPremier Base Color must be activated with ChromaPremier Activator 12305S before use. Mix 1 part ChromaPremier Base Color with 1 part ChromaSystem™ Basemaker®. Stir thoroughly, then activate: add 1 ounce 12305S Activator to 1 ready-to-spray quart of Base Color ($^1/_2$ ounce to 1 ready-to-spray pint). Spray viscosity is 15-17 seconds in a #2 Zahn.

POT LIFE: 8 hours.

APPLICATION: Apply Base Color in two to three medium coats until desired hiding and appearance are achieved. Flash 10-15 minutes before clear. See Technical Manual for further information.

FOR VOC REGULATED AREAS: These directions refer to the use of products which may be restricted or require special mixing instructions in VOC regulated areas. Follow mixing and usage recommendations in the VOC Compliant Products Chart for your area.

See Individual Label Directions for Use of the Other DuPont Products

Made in U.S.A.

DUPONT AUTOMOTIVE • REFINISH PRODUCTS • Wilmington, Delaware 19898
For medical & environmental information: (800) 441-7515

4RF-065ABAC-080-0996
E-R1306 H-42251

FIGURE 11–2 *Paint can label from a three-part basecoat. Note the mixing ratio and viscosity. (Courtesy of DuPont Automotive Refinish Products.)*

in a 4:2:1 reduction, four parts of the first component are combined with two parts of the second component and with one part of the last component. If one-half pint is determined to be the part size, this ratio would be two pints to one pint to one-half pint. The volume of the mixed paint would be three-and-one-half pints. Another way to obtain this ratio is to decide on the amount of sprayable paint needed. Obtain an appropriate container. Place a paint stick into the container and mark on the stick the desired height of the mixed paint. Remove the stick from the can. Measure the distance from the bottom of the stick to the height mark. Divide that distance into seven equal parts (4 + 2 + 1 = 7). Mark the seven parts on the stick. Put the marked stick into the container and add the proper amount of each component.

A percentage mixture uses the main component, usually paint, as the standard measure. The additives are listed as a percentage of the main component. For example, if a 50% reduction is needed and there is one pint of paint to reduce, one-half pint of reducer (50% or half of the volume of the standard) is added to the paint (Figure 11–3). A 50% reduction can be stated as a ratio of 2:1.

The combination ratio and percentage are cumulative. In a dilution of 4:2:30%, two parts of the second

component are added to four parts of the first. Then the volume of this 4:2 mixture is increased by 30% with the last component. For example, if the volume of the 4:2 mixture is one pint, then slightly less than one-third of one pint of the last component is added to the mixture.

SHOP TALK

Always thoroughly stir base coats, color, or clear coat before pouring or measuring. Reducers should be shaken by hand before mixing.

The easiest way to obtain specified portions of components is to use a mixing stick (Figure 11–4). The use of a mixing stick will avoid proportion problems. The mixing stick is marked off in proper proportions for different amounts of sprayable paint (Figure 11–5). In a three-component mixing stick, there are three ones, three twos, three threes, and so on. First obtain the proper sized container, and put the mixing stick in. Decide on the amount of sprayable paint needed. For example, if the amount of sprayable paint needed is level with the upper or right-hand one, add the first component to the level of the first one, the second component to the level of the second one,

Reduction / thinning percentage		Reduction proportions	Paint		Solvent	
10%	=	5 parts paint / 1 part solvent		10%		
25%	=	4 parts paint / 1 part solvent		25%		
33%	=	3 parts paint / 1 part solvent		33%		
50%	=	2 parts paint / 1 part solvent		50%		
75%	=	4 parts paint / 3 parts solvent		75%		
100%	=	1 part paint / 1 part solvent		100%		
125%	=	4 parts paint / 5 parts solvent		125%		
150%	=	2 parts paint / 3 parts solvent		150%		
200%	=	1 part paint / 2 parts solvent		200%		
250%	=	2 parts paint / 5 parts solvent		250%		

FIGURE 11–3   Percentage reduction chart.

FIGURE 11–4   Paint mixing stick in use.

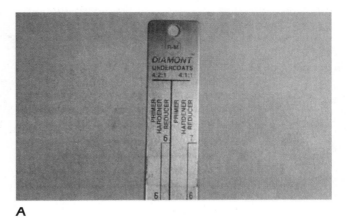

A

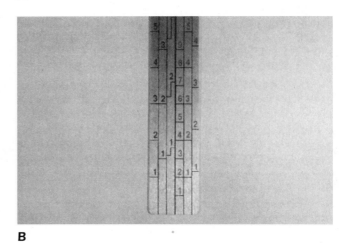

B

FIGURE 11–5   Paint mixing stick graduations.

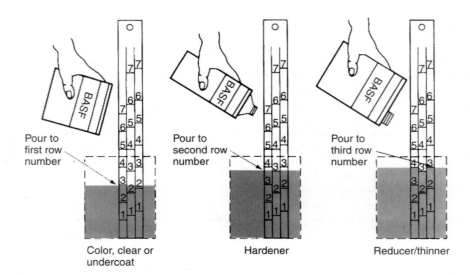

Pour to first row number — Color, clear or undercoat

Pour to second row number — Hardener

Pour to third row number — Reducer/thinner

FIGURE 11–6 *How to use a mixing stick.*

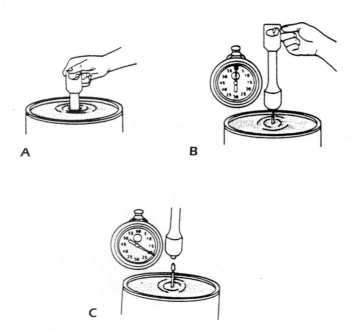

A

B

C

FIGURE 11–7 *How to use a #2 Zahn cup to measure paint viscosity.*

and the final component to the level of the last one (Figure 11–6).

Another way to mix paint is to use graduated disposable cups. These cups are marked off in different ratios and percentages. Simply find the ratio that applies to the paint and add the components. The cup can be thrown away after use.

Some shops use graduated glass or stainless steel cups for mixing. These can be reused. However, with glass cups, there is always a chance of breakage. The cups can be rinsed in the spray gun cleaner after use.

Once the paint is mixed, the viscosity can be checked. The easiest way to check viscosity is to use a #2 Zahn cup (Figure 11–7). The Zahn cup is a cylinder with a hole in the bottom. The viscosity of paint is measured by how many seconds it takes for the full cup to empty.

To determine paint viscosity with a #2 Zahn cup, proceed as follows:

1. Prepare the material to be tested. Mix, strain, and reduce as directed by the manufacturer.
2. Fill the cup by submerging it in the material.
3. Release the flow of the material and trigger the stopwatch. Keep all eyes on the flow, not on the watch.
4. When the solid stream of material "breaks" (indicating air passing through the orifice), stop the watch.
5. The result is expressed in seconds.

If the viscosity measurement of a two-part paint is not within specifications, more paint or reducer may be added to correct the problem. For example, if the specification is 18–22 seconds and the mixed paint viscosity is 16 seconds, the mixed paint has too much reducer because the mixed paint ran through the Zahn cup faster than the specified time. To correct the problem, paint should be added. On the other hand, if the mixed paint runs through the Zahn cup too slowly (24 seconds), the paint is too thick. Add reducer and retest. This checking and addition works only for a two-part paint. If the viscosity of a three-part paint is off it is best to discard the mixed paint and remix. The portions are critical; adding more of one component will throw the ratio off.

Once the paint is mixed, it must be filtered before it is added to the spray gun cup (Figure 11–8). The filter

FIGURE 11–8 *Filter the paint as it is added to the paint gun.*

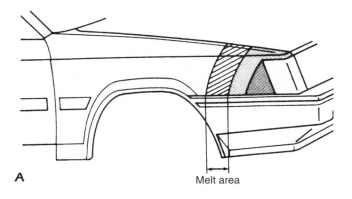

A
Melt area

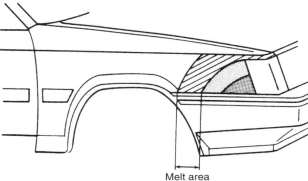

B
Melt area

mesh traps contaminates in the paint, preventing them from entering the gun cup. Paint filters can be disposable (made out of paper) or reusable (made out of plastic).

SHOP TALK

The temperature of the vehicle and the paint are important in obtaining a good paint job. Both the vehicle and the paint should be at room temperature, ideally 70°F. Cold paint sprayed on a cold car will run. In cold climates, make sure the vehicle and the paint are at room temperature before spraying.

FIGURE 11–9 *(A) Single-stage spot paint, entire fender. (B) Single-stage spot paint, partial fender. Body line is backtaped. This method gives a good match between the hood and the fender.*

11.2 SINGLE-STAGE PAINT

Vehicles with OEM single-stage paint can be refinished with the following paints:

• Acrylic enamel with hardener
• Acrylic urethane

The acrylic urethane single-stage is more durable than acrylic enamel with hardener. However, either system should give at least five years of durability. The single-stage paint is available in solid or metallic.

SPOT PAINT

Spot painting means to paint a partial panel (Figure 11–9). Because refinish overspray must be melted into the existing paint, this method has limited durabil-

ity due to thin paint film build. It can be used if durability is not an issue or if a refinished single-stage panel is scratched during reassembly.

The **melt** area should be cleaned with silicon free rubbing compound. The paint area should be sanded with 600-grit sandpaper (Figure 11–10).

Mix the paint; load it into the spray gun and adjust the spray gun. A second spray gun is to be loaded with enamel reducer. This is the two-gun method. The dry edge around the repair is dusted with reducer from the second spray gun to melt in the overspray. The first coat of paint is sprayed over the primer. Arcing is allowed to minimize the overspray on solid colors (Figure 11–11). The second gun is set to about 25 psi air pressure, and the overspray area is dusted with reducer. This is a delicate operation. Too much reducer will run. After the flash time is observed, the next coat of paint is brought out an inch beyond the first. The overspray is dusted with the second spray gun. The final coat is applied, after the flash time, in the rubbed area. The overspray is melted.

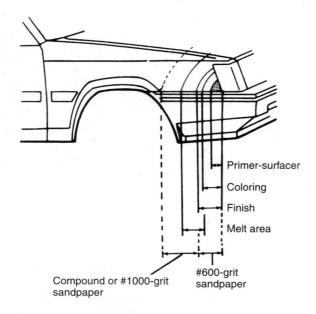

FIGURE 11-10   How to prep for spot paint.

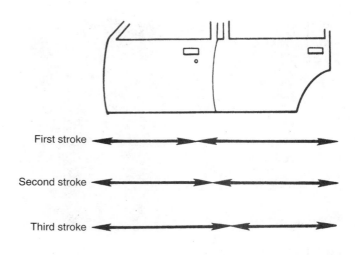

FIGURE 11-12   Changing overlap areas.

FIGURE 11-11   Arcing of spray gun when spot painting.

SHOP TALK

An OEM single-stage paint may be available in refinish basecoat/clear coat. Basecoat/clear coat is much easier to blend than single-stage paint.

PANEL PAINT

Panel painting means to paint a part with a definite boundary, such as a door or a fender. If the color match is not a problem, this method works well. Simply prep the panel as explained in Chapter 9 and mask off. Mix the paint following the manufacturer's recommendations. Load the diluted paint into the gun and adjust the gun. Tack off the panel. Spray the first coat. Allow the paint to flash. Spray a wet coat. Allow the paint to flash. Spray on another wet coat.

OVERALL PAINTING

When overall spraying, the painter can keep a wet edge while maintaining minimum overspray on the horizontal surfaces. This prevents spray dust from settling into areas that have already dried, which would cause a gritty surface. Avoid sags in the overlap line by changing the point of overlapping as shown in Figure 11-12.

Although there is not a single most perfect procedure for repainting a car overall, most refinishers will agree that the diagrams in Figure 11-12 illustrate the best patterns. With a crossdraft booth, by starting with the top of the car and proceeding to the trunk deck lid, side, and so on, the painter can best keep a "wet edge" while maintaining minimum overspray on the horizontal surfaces (Figure 11-13). This prevents spray dust from settling onto areas that have already dried, causing a gritty or contaminated surface. If possible, it is better for two painters to work in a state-of-the-art downdraft booth. The spray pattern is different than that of the conventional booth because of the direction of the airflow (top to bottom). Following the pattern shown in Figure 11-14 allows the three main horizontal surfaces to remain as wet as possible while maintaining minimum overspray. These procedures also allow the painter(s) to continue to apply additional coats as needed without a significant loss of time due to flash off between coats.

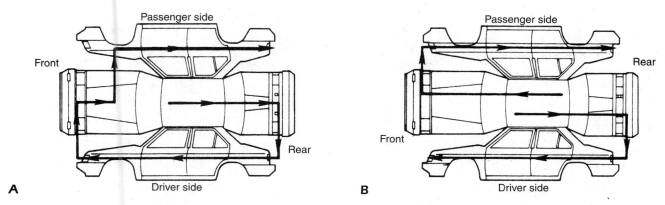

FIGURE 11–13 *Overall painting procedures in a crossdraft booth: painting order for (A) one painter and (B) two painters.*

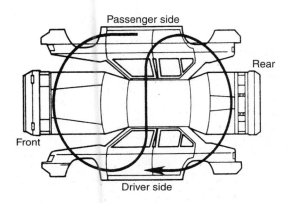

FIGURE 11–14 Painting procedure in a downdraft booth.

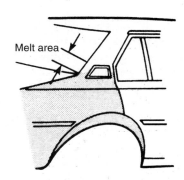

FIGURE 11–15 *Melt area on a sail panel.*

11.3 — TWO-STAGE PAINT

Two-stage or basecoat/clear coat is by far the most commonly used method of refinishing vehicles. Blending is easy. The clear coat is very durable, allowing collision repair facilities to warrant repairs for as long as the owner has the vehicle.

SPOT PAINT

Spot painting of basecoat/clear coat is not recommended except in the melt of clear coat in a sail panel (Figure 11–15). This area is usually narrow, less than twelve inches. Do not attempt a melt in a broad area.

In this example, the basecoat is applied on the quarter panel and the refinish clear is melted into the sail panel clear coat. To make this repair, follow the surface prep instructions in Chapter 9. If adhesion promoter is needed, spray it beyond the melt area. Be careful; this clear product runs easily. When the adhesion promoter has dried, spray the basecoat. Step each basecoat out about one to two inches from the previous coat (Figure 11–16). Allow proper flash time. Mix the clear coat and load the spray gun. Load another gun with blending clear. If blending clear is not available, a mixture of clear and slow dry reducer (one part clear and hardener and ten parts reducer) can be made. Spray a coat of clear over all of the basecoat. Melt in the clear overspray with the blender. Use about 25 psi of air pressure. Be careful; a dusting is all that is required. Too much will run. Spray on another coat of clear over all of the basecoat. Extend out about one inch from the edge of the previous

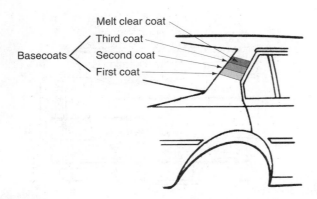

FIGURE 11-16 *Blending basecoat and melting clear coat on a sail panel.*

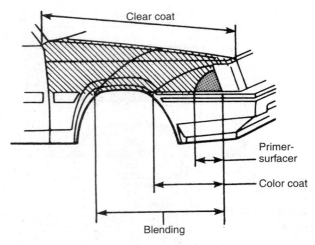

FIGURE 11-17 *Blending on a fender. Some shops may choose to blend the bumper, remove the fender rear moulding, and clear coat the entire fender.*

clear. Melt in with the blender. This type of repair is not durable at the melt because of insufficient film build. However, the only alternative that is durable in this quarter panel example is to paint the quarter and clear coat the quarter, roof, and opposite quarter panel. This may be required by some vehicle manufacturers in order to maintain the factory paint warranty.

PANEL PAINT

Panel painting is the preferred method of refinishing basecoat/clear coat. Blending the color is easy. The entire panel is clear coated, so the finish is durable with proper film build. In a fender repair (Figure 11-17), the following steps are used:

1. The fender is prepped, following the instructions in Chapter 9.
2. Basecoat is applied. Tack off overspray between each coat of base. There are two ways to blend:

• **Front blend**. The first coat covers the primer only. Each successive coat extends out about one to two inches (Figure 11-18).
• **Back blend**. The first coat of color covers the primer and extends out to the edge of the blend. Each successive coat covers a smaller area inside the first coat (Figure 11-19). Avoid arcing the gun when spraying a metallic basecoat. The dry spray from the arc will show up as a bright ring at the blend area. This is sometimes called the *halo effect*. Make the entire coat of base the same wetness.

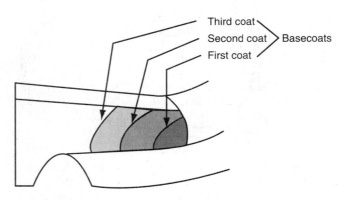

FIGURE 11-18 *Front blend, entire fender is clear coated.*

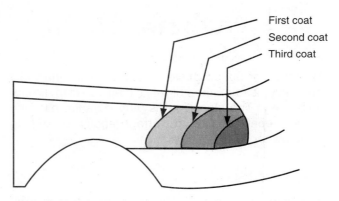

FIGURE 11-19 *Back blend, entire fender is clear coated.*

Transparent colors, such as red and red pearl, may require more than three coats of base to hide a light colored surfacer or sealer. If hiding is expected to be a problem, tint the sealer to match the basecoat.

3. Spray two to three coats of clear over the entire fender.

This panel painting can include multiple panels as in a fender replacement. In this case, the fender is painted and the hood, bumper (if painted), and door are blended. Usually plan on clear coating the entire blended panel. Do not try to melt the clear unless the melt area is narrow—less than twelve inches—and will save considerable paint material and labor time. Remember that a blended panel that is not entirely clear coated will not be durable.

COMPLETE

A basecoat/clear coat "complete" is not much different from spraying a multiple panel basecoat/clear coat. Follow the pattern shown in Figures 11–13 or 11–14. Allow the proper **flash time** between basecoats. Tack off between basecoats and before the first coat of clear.

 **THREE-STAGE PAINT**

Three-stage paints, or tricoats, consist of a solid base color, a pearl midcoat, and a clear coat. The number of pearl coats is critical in color matching. Chapter 13 explains how to make a let-down panel to check on the required number of midcoats. The essential difference between two-stage and three-stage painting is only the number of coats and the amount of area for blending.

SPOT PAINT

Spot painting of tricoats, where the clear coat would be melted all around the repair, is not recommended, except that the clear coat may be melted in the same manner as the clear on the sail panel of a basecoat/clear coat. If the repair area is small (confined to a single panel), the preferred repair method would be panel painting: blend the base color and midcoat, then clear coat the entire panel.

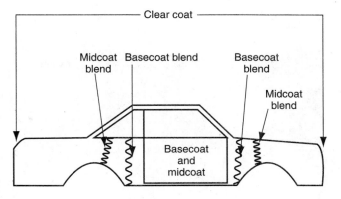

FIGURE 11–20 Tricoat blend.

PANEL PAINT

Figure 11–20 shows an example of a tricoat vehicle with a door replacement. In this example, the door, fender, and quarter panel are prepped. The door is basecoated. Check the manufacturer's procedures. The blend areas may require overreduction. The basecoat is blended onto the fender and quarter panel. The midcoat is sprayed to cover the basecoat. Then the midcoat is blended further onto the fender and quarter panel. The front blend method is recommended. Step out each midcoat about six inches from the previous coat. Do not arc the gun. The entire side of the vehicle is clear coated.

This process can also be used on a single panel such as a fender. However, because of the amount of area required for blending the midcoat, it may not always be possible to confine the blending to one panel.

COMPLETE

A tricoat complete consists of prepping the entire vehicle and spraying the base, midcoat, and clear coats.

 **TWO-TONE PAINT**

Two-tone vehicles have a dark color and a light color. Pickup trucks are the most common vehicles found with two-tones (Figure 11–21). Single-stage and BC/CC two-tones are set up in similar ways.

- **Single-stage two-tone.** Prep the surface for topcoat application. There are two ways to mask,

FIGURE 11-21   Two-tone pickup truck.

so two colors can be sprayed. The first method is to spray one color overlapping the area that will be sprayed with the contrasting color. When the first color has dried, at least one hour, tape off the first color with fine-line tape. Mask off the first color. Sand the overspray and the area around the fine-line tape. Spray the second color. Remove the masking paper and tape.

The other method is to mask off one of the areas with fine-line tape and paper. Spray the first color. Remove the masking and allow the paint to dry. Mask off the first color with fine-line taping at the paint line. Be sure that there aren't any gaps between the tape and paint. Mask off the first color. Spray the second color. Remove the masking.

- **Basecoat/clear coat two-tone.** Prep the surface for topcoat application. The masking method can follow either of the procedures explained in the single-stage two-tone. Spray on one color. It can be blended if required. The second color is sprayed after a sixty-minute dry time. Remove the masking. The entire refinish area is sprayed with two to three coats of clear.

REVIEW QUESTIONS

1. Technician A says that paint viscosity is measured with a Ford cup. Technician B says that paint viscosity is measured with a Zahn cup. Who is right?
   a. technician A
   b. technician B
   c. both technician A and technician B
   d. neither technician A nor technician B
2. Which paint is most durable?
   a. acrylic enamel
   b. acrylic urethane
   c. acrylic enamel with hardener
3. Technician A says that spot painting is not as durable as panel painting. Technician B says that proper paint build is needed for durability. Who is right?
   a. technician A
   b. technician B
   c. both technician A and technician B
   d. neither technician A nor technician B
4. When blending a metallic basecoat technician A arcs the gun at the blend edge. Technician B does not arc the gun at the blend edge. Who is right?
   a. technician A
   b. technician B
   c. both technician A and technician B
   d. neither technician A nor technician B
5. Technician A says that urethane basecoat/clear coat paint may be warranted for as long as the owner has the vehicle. Technician B says that a melt area should not be wider than twelve inches. Who is right?
   a. technician A
   b. technician B
   c. both technician A and technician B
   d. neither technician A nor technician B
6. The bright ring around a metallic blend is caused by
   a. uniform wet spray.
   b. arcing the spray gun.
7. Technician A says that a blend can hide a slight color mismatch. Technician B plans on blending whenever a panel is painted. Who is right?
   a. technician A
   b. technician B
   c. both technician A and technician B
   d. neither technician A nor technician B
8. A _____ panel is made to match a tricoat finish.
9. Technician A tacks off the surface before each basecoat. Technician B says that a melt on a sail panel is not warranted. Who is right?
   a. technician A
   b. technician B
   c. both technician A and technician B
   d. neither technician A nor technician B
10. When one is spraying a basecoat/clear coat two-tone vehicle, the first basecoat is applied and clear coated. Then the second basecoat is sprayed and cleared.
    a. True
    b. False

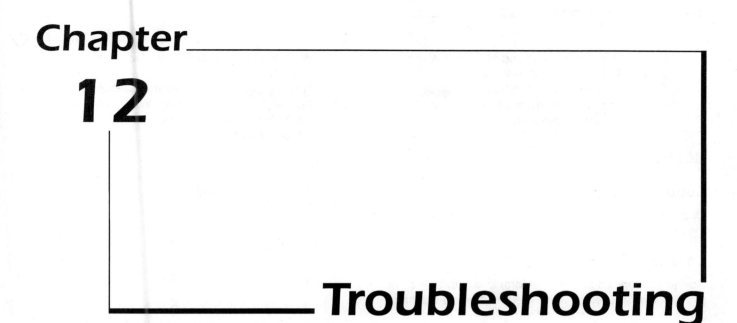

Troubleshooting

Objectives

After reading this chapter, you will be able to recognize and correct defects that occur in a paint finish.

12.1 — EXISTING PROBLEMS

FADING

Condition

Gradual color change due to pigment exposure to sunlight. All pigments change due to sunlight exposure. Some, such as red or yellow, change rapidly. Others, such as blue and green, change slowly.

Causes

Pigment exposure to sunlight.

Prevention

Clear coat reduces the rate of fading.

Solution

1. Color-sand and buff off the upper, faded paint.
2. Sand and refinish in extreme cases.

ACID RAIN

Condition

Spotty discoloration that appears on the surface. (Various pigments react differently when in contact with acid or alkalies.)

Causes

Chemical change of pigments resulting from atmospheric contamination, in the presence of moisture (acid rain), due to industrial activity.

Prevention

1. Keep the finish away from a contaminated atmosphere.
2. Immediately following contamination, the surface should be vigorously flushed with cool water and detergent.

Solution

1. Wash with detergent water and follow with a vinegar bath.

2. If alkaline, wash with baking soda and water. If acid, buff to remove the staining.
3. If contamination has reached the metal or subcoating, the spot must be sanded down to the metal before refinishing.

BLEEDING

Condition

Original finish discoloring—or color seeping through—the new topcoat color.

Causes

Contamination, usually in the form of soluble dyes or pigments on the older finish before it was repainted. (This is especially true with older shades of red.) Hardener in polyester filler can also cause a yellowing in basecoat/clear coat finishes.

Prevention

Thoroughly clean areas to be painted before sanding, especially when applying light colors over dark colors. (Avoid using light colors over older shades of red without sealing first.) To prevent bleed-through, apply a coat of nonstaining filler over the filler.

Solution

Sand and prep the surface. Apply two medium coats of sealer. Then reapply the color coat.

CHALKING

Condition

This occurs when pigment powder is no longer held by the binder. It makes the finish look dull.

Causes (Other Than Normal Exposure)

1. Wrong thinner or reducer, which can harm topcoat durability.
2. Materials not uniformly mixed.
3. Thin paint film.
4. Excessive mist coats when finishing a metallic color application.

Prevention

1. Select the thinner or reducer that is best suited for existing shop conditions.
2. Stir all pigmented undercoats and topcoats thoroughly.

3. Meet or slightly exceed minimum film thicknesses.
4. Apply metallic color as evenly as possible so that misting is not required. When mist coats are necessary to even out flake, avoid using straight reducer.

Solution

Remove surface in affected area by sanding, then clean and finish.

CHIPPING

Condition

Small chips of a finish losing adhesion to the substrate, usually caused by the impact of stones or hard objects. While refinishers have no control over local road conditions—and thus cannot prevent such occurrences—they can take steps to minimize the effects if they know beforehand that these conditions will exist. (For details on the causes, prevention, and solution for chipping, see *Peeling* later in this chapter.)

CRACKING

Condition

A series of deep cracks resembling mud cracks in a dry pond. Often occurring in the form of three-legged stars or in no definite pattern, they are usually through the color coat and sometimes the undercoat as well.

Causes

1. Excessive film thickness. Excessively thick topcoats magnify normal stresses and strains that can result in cracking even under normal conditions.
2. Materials not uniformly mixed.
3. Insufficient flash time.
4. Incorrect use of additive.

Prevention

1. Do not pile on topcoats. Allow sufficient flash and dry time between coats. Do not dry by gun fanning.
2. Stir all pigmented undercoats and topcoats thoroughly. Strain and—where necessary—add fisheye eliminator to topcoats.
3. Same as Step 1.

4. Read and carefully follow label instructions. Additives not specifically designed for a color coat can weaken the final paint film and make it more sensitive to cracking.

Solution

The affected areas must be sanded to a smooth finish or, in extreme cases, removed down to the bare metal and refinished.

LINE CHECKING

Condition

Similar to cracking, except that the lines or cracks are more parallel and range from very short to about eighteen inches.

Causes

1. Excessive film thickness.
2. Improper surface preparation. Oftentimes occurs with the application of a new finish over an old film that had cracked and was not completely removed.

Prevention

1. Do not pile on topcoats. Allow sufficient flash and dry time. Do not dry by gun fanning.
2. Thoroughly clean areas to be painted before sanding. Be sure the surface is completely dry before applying any undercoats or topcoats.

Solution

Remove color coat down to primer and apply new color coat.

MICROCHECKING

Condition

Appears as severe dulling of the film, but when examined with a magnifying glass, it contains many small cracks that do not touch. Microchecking is the beginning of film breakdown and might be an indication that film failures such as cracking or crazing will develop.

Solution

Sand off the color coat to remove the cracks, then recoat as required.

CRAZING

Condition

Fine splits or small cracks—often called "crow's feet"—that completely checker an area in an irregular manner.

Causes

Shop too cold. Surface tension of original material is under stress and literally shatters under the softening action of the solvents being applied.

Prevention

Select the thinner or reducer that is suitable for existing shop conditions. Schedule painting to avoid temperature and humidity extremes in the shop or paint at a temperature between that of the shop and that of the job. Bring the vehicle to room temperature before refinishing.

Solution

1. Continue to apply wet coats of topcoat to melt the crazing and flow the pattern together (using the wettest possible reducer that shop conditions will allow).
2. Use a fast-flashing reducer, which will allow a bridging of subsequent topcoats over the crazing area. (This is one case where bridging is a cure and not a cause for trouble.)

DULLED FINISH

Condition

Gloss retards as film dries.

Causes

1. Compounding before reducer evaporates.
2. Using poorly balanced reducer.
3. Poorly cleaned surface.
4. Topcoats put on wet subcoats.
5. Washing with caustic cleaners.
6. Inferior polishes.
7. Improper air flow in spray bottle.
8. Insufficient flash time.

Prevention

1. Clean the surface thoroughly.
2. Use recommended materials.
3. Allow all coatings sufficient drying time.

Solution

Allow finish to dry hard and rub with a mild rubbing compound.

PEELING

Condition

Loss of adhesion between paint and substrate (top-coat to primer and/or old finish, or primer to metal).

Causes

1. Improper cleaning or preparation. Failure to remove sanding dust and other surface contaminants will keep the finish coat from coming into proper contact with the substrate.
2. Improper metal treatment.
3. Topcoat, primer, surfacer or sealer not properly mixed.
4. Failure to use proper sealer.

Prevention

1. Thoroughly clean areas to be painted. It is always good shop practice to wash the sanding dust off the area to be refinished with cleanup solvent.
2. Use correct metal conditioner and conversion coating.
3. Stir all pigmented undercoats and topcoats thoroughly.
4. In general, sealers are recommended to improve the adhesion of topcoats.

Solution

Remove finish from an area slightly larger than the affected area and refinish.

RUST UNDER FINISH

Condition

The surface will show raised surface spots or peeling or blistering.

Causes

1. Improper metal preparation.
2. Broken paint film allows moisture to creep under surrounding finish.
3. Water in air lines.

Prevention

1. Locate the source of moisture and seal it off.

2. When replacing ornaments or mouldings, be careful not to break the paint film and allow dissimilar metals to come into contact. This contact can produce electrolysis that might cause a tearing away or loss of good bond with the paint film.

Solution

1. Seal the inner parts of panels from moisture.
2. Sand down to bare metal, prepare metal, and treat with phosphate before refinishing.

STONE BRUISES

Condition

Small chips of paint missing from an otherwise firm finish.

Causes

1. Flying stones from other vehicles.
2. Impact of other car doors in a parking lot.

Solution

1. Thoroughly sand remaining paint film back several inches from damage point.
2. Properly treat metal and refinish.

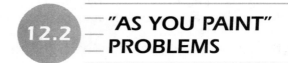

12.2 "AS YOU PAINT" PROBLEMS

WATER SPOTTING

Condition

General dulling of gloss in spots or masses of spots.

Causes

1. Water evaporating on finish before it is thoroughly dry.
2. Washing finish in bright sunlight.

Prevention

1. Do not apply water to a fresh paint job, and try to keep a newly finished car out of the rain or dew. Allow sufficient dry time before delivering the car to the customer.
2. Wash the car in shade and wipe it dry.

Solution

Use rubbing or polishing compound. In severe cases, sand affected areas and refinish.

WET SPOTS

Condition

Discoloration and/or slow drying of some areas.

Causes

1. Improper cleaning and preparation.
2. Improper drying of excessive undercoat film build.
3. Sanding with contaminated solvent.

Prevention

1. Thoroughly clean all areas to be painted.
2. Allow proper drying time for undercoats.
3. Wet-sand with clean water.

Solution

Wash or sand all affected areas thoroughly and then refinish.

WRINKLING

Condition

Surface distortions (or shriveling) that occur while the enamel topcoat is being applied (or later during the drying stage).

Causes

1. Improper dry. When a freshly applied topcoat is baked or force-dried too soon, softening of the undercoats can occur. This increases topcoat solvent penetration and swelling. In addition, baking or force-drying causes surface layers to dry too soon. The combination of these forces causes wrinkling.
2. "Piling on" heavy or wet coats. When enamel coats are too thick, the lower wet coats are not able to release their solvents and set up at the same rate as the surface layer, which results in wrinkling.
3. Improper reducer or incompatible materials. A fast-dry reducer or the use of a lacquer thinner in enamel can cause wrinkling.

4. Improper or rapid change in shop temperature. Drafts of warm air cause enamel surfaces to set up and shrink before sublayers have released their solvents, which results in localized skinning in uneven patterns.

Prevention

1. Allow proper drying time for undercoats and topcoats. Read and carefully follow label instructions.
2. Do not pile on topcoats. Allow sufficient flash and dry time.
3. Select proper reducer and avoid using incompatible materials such as a reducer with lacquer products or thinner with enamel products.
4. Schedule painting to avoid temperature extremes or rapid changes.

Solution

Remove wrinkled enamel and refinish.

BLUSHING

Condition

A milky white haze that appears on paint films.

Causes

1. In hot humid weather, moisture droplets become trapped in the wet paint film. Air currents from the spray gun and the evaporation of the reducer tend to make the surface being sprayed lower in temperature than the surrounding atmosphere. This causes moisture in the air to condense on the wet paint film.
2. Excessive air pressure.
3. Too fast a reducer.

Prevention

1. In hot, humid weather, try to schedule painting early in the morning when temperature and humidity conditions are more suitable.
2. Use proper gun adjustments and techniques.
3. Select the reducer that is suitable for existing shop conditions.

Solution

Add retarder to the reduced color and apply additional coats.

DIRT IN FINISH

Condition

Foreign particles dried in the paint film.

Causes

1. Improper cleaning, blowing off, and tack-ragging of the surface to be painted.
2. Defective air regulator cleaning filter.
3. Dirty working area.
4. Defective or dirty air inlet filters.
5. Dirty spray gun.
6. Dirty painter.

Prevention

1. Blow out all cracks and body joints.
2. Solvent-clean and tack-rag surface thoroughly.
3. Be sure equipment is clean.
4. Work in clean spray area.
5. Replace inlet air filters if dirty or defective.
6. Strain out foreign matter from paint.
7. Keep all containers closed when not in use to prevent contamination.
8. Wear a clean paint suit.

Solution

1. Rub out the finish with compounds.
2. If dirt is deep in the finish, sand and compound to restore gloss. Metallic finishes might show mottling with this treatment and will then require additional color coats.

FEATHEREDGE SPLITTING

Condition

Appears as stretch marks (or cracking) along the featheredge. Occurs during or shortly after the top-coat is applied over lacquer surfacer.

Causes

1. "Piling on" the undercoat in heavy and wet coats. Solvent is trapped in undercoat layers that have not had sufficient time to set up.
2. Material not uniformly mixed. Because of the high pigment content of surfacers, it is possible for settling to occur after they have been thinned. Delayed use of this material without restirring results in applying a film with loosely held pigment containing voids and crevices throughout, causing the film to act like a sponge.
3. Wrong thinner.

4. Improper surface cleaning or preparation. When not properly cleaned, surfacer coats can draw away from the edge because of poor wetting and adhesion.
5. Improper drying. Fanning with a spray gun after the surfacer is applied will result in drying the surface before solvent or air from the lower layers is released.
6. Excessive use (and film build) of putty.

Prevention

1. Apply properly reduced surfacer in thin to medium coats with enough time between coats to allow solvents and air to escape.
2. Stir all pigmented undercoats and topcoats thoroughly. Select thinner that is suitable for existing shop conditions.
3. Select only thinners that are recommended for existing shop conditions.
4. Thoroughly clean areas to be painted before sanding.
5. Apply surfacer in thin to medium coats with enough time between coats to allow solvents and air to escape.
6. Lacquer putty should be limited to filling minor imperfections. Putty applied too heavily (or too thick) will eventually shrink, causing featheredge splitting.

Solution

Remove finish from the affected areas and refinish.

FISHEYES

Condition

Small, crater-like openings in the finish after it has been applied.

Causes

1. Improper surface cleaning or preparation. Many waxes and polishes contain silicone, the most common cause of fisheyes. Silicones adhere firmly to the paint film and require extra effort to remove. Even small quantities in sanding dust, rags, or from cars being polished nearby can cause this failure.
2. Effects of the old finish or previous repair. Old finishes or previous repairs can contain excessive amounts of silicone from additives used during their application. Usually solvent wiping will not remove embedded silicone.
3. Contamination of air lines.

Prevention

1. Precautions should be taken to remove all traces of silicone by thoroughly cleaning with wax and grease solvent. The use of fisheye eliminator is in no way a replacement for good surface preparation.
2. Add fisheye eliminator to the paint. The fisheye eliminator contains silicon. The paint will not be repelled from the contaminated area. The paint will be able to flow out and cover the fisheye.
3. Drain and clean the air pressure regulator daily to remove trapped moisture and dirt. The air compressor tank should also be drained daily.

Solution

Alllow the paint to flash. Wipe off the area with mild wax and grease remover. Sand with 1,000-grit sandpaper and water. Be careful not to roll up the paint edge. Rewipe with mild wax and grease remover. Apply a dry coat of paint over the sanded area.

Fisheye eliminator may be added to the paint, following manufacturers' recommendations. Avoid the use of fisheye eliminator by proper surface preparation.

LIFTING

Condition

Surface distortion or shriveling while the topcoat is being applied or while drying.

Causes

1. Use of incompatible materials. Solvents in the new topcoat attack the old surface, which results in a distorted or wrinkled effect.
2. Insufficient flash time. The solvents from the coat being applied cause localized swelling or partial dissolving that later distorts the final surface.
3. Effect of old finish or previous repair. Enamel sprayed over a previous refinish lacquer surfacer may cause lifting. The solvent in the enamel causes the lacquer to dissolve. Usually the lifting is at the feather edge.
4. Improper surface cleaning or preparation.

Prevention

1. Avoid incompatible materials, such as a thinner with enamel products, or incompatible sealers and primers.

2. Do not pile on topcoats. Allow sufficient flash and dry time. The final topcoat should be applied when the previous coat is still soluble or after it has completely dried and is impervious to topcoat solvents.
3. Properly clean surfaces prior to painting.

Solution

Allow the lifted area to dry for twenty minutes. Lightly sand with 600-grit sandpaper and water. Sand only on the surface; do not dig in. Sand until the surface is smooth. Clean off the water. Adjust the spray gun for dry spray by turning in the fluid adjustment knob. The purpose of dry spray is to build up an impervious layer of dry paint. Dust a coat of paint onto the lifted area. Allow two minutes flash time, tack off the dry overspray, and dust another coat on. After flash time and tacking, spray a third dry coat. The area is now ready for topcoat. Allow longer-than-normal flash times between coats.

In extreme cases, a lifted area should be allowed to dry. After drying, the area is sanded smooth. A water-based sealer is sprayed on. The water-based sealer, because it does not contain reducer or thinner, will not make the area lift. Allow the sealer to dry twenty-four hours. Lightly sand and then topcoat. The water-based sealer prevents the solvents in the enamel from contacting the lacquer surfacer.

MOTTLING

Condition

Occurs only in metallics when the flakes float together to form a spotty or striped appearance.

Causes

1. Wrong reducer.
2. Materials not uniformly mixed.
3. Spraying too wet.
4. Holding spray gun too close to work.
5. Uneven spray pattern.
6. Low shop temperature.

Prevention

1. Select the reducer that is suitable for existing shop conditions, and mix properly. In cold, damp weather use a faster-drying solvent.
2. Stir all pigmented topcoats—especially metallics—thoroughly.
3. Use proper gun adjustments, techniques, and air pressure.

4. Keep your spray gun clean (especially the needle fluid tip and air cap) and in good working condition.

Solution

Allow the color coat to set up. Clean the spray gun. Spray at right angles to the stripes.

ORANGE PEEL

Condition

Uneven surface formation—much like that of the skin of an orange—that results from poor fusion of atomized paint droplets. Paint droplets dry out before they can flow out and level smoothly together.

Causes

1. Improper gun adjustment and techniques. Too little or too much air pressure, wide fan patterns, or spraying at excessive gun distances cause droplets to become too dry during their travel time to the work surface, and they remain as formed by the gun nozzle.
2. Extreme shop temperature. When the air temperature is too high, droplets lose more solvent and dry out before they can flow and level properly.
3. Improper dry. Gun fanning before paint droplets have a chance to flow together will cause orange peel.
4. Improper flash or recoat time between coats. If the first coats of enamel are allowed to become too dry, solvent in the paint droplets of following coats will be absorbed into the first coat before proper flow is achieved.
5. Wrong reducer. Underdiluted paint or paint reduced with a fast-evaporating reducer causes the atomized droplets to become too dry before reaching the surface.
6. Too little reducer.
7. Materials not uniformly mixed. Many finishes are formulated with components that aid fusion. If these are not properly mixed, orange peel will result.

Prevention

1. Use proper gun adjustments, techniques, and air pressure.
2. Schedule painting to avoid temperature and humidity extremes. Select the reducer that is suitable for existing conditions. The use of a slower-evaporating reducer will overcome extreme temperatures.

3. Allow sufficient flash and dry time. Do not dry by fanning.
4. Allow proper drying time for undercoats and topcoats (neither too long nor too short).
5. Select the reducer that is most suitable for existing shop conditions to provide good flow and leveling of the topcoat.
6. Reduce to recommended viscosity with proper reducer.
7. Stir all pigmented undercoats and topcoats thoroughly.

Solution

Adjust the spray gun, increasing air pressure by five pounds. If the paint is cured, it may be color-sanded and buffed.

PINHOLES

Condition

Tiny holes or groups of holes in the finish, or in putty or body filler, usually are the result of trapped solvents, air, or moisture.

Causes

1. Improper surface cleaning or preparation. Moisture left on surfacers will pass through the wet topcoat to cause pinholing.
2. Contamination of air lines. Moisture or oil in air lines will enter paint while it is being applied and will cause pinholes during the drying stage.
3. Wrong gun adjustment or technique. If adjustments or techniques result in an application that is too wet, or if the gun is held too close to the surface, pinholes will occur when the air or excessive solvent is released during drying.
4. Wrong reducer. The use of a solvent that is too fast for the shop temperature tends to make the refinisher spray too close to the surface in order to get adequate flow. When the solvent is too slow, it is trapped by subsequent topcoats.
5. Improper dry. Fanning a newly applied finish can drive air into the surface or cause a dry skin, both of which result in pinholing when solvents retained in lower layers come to the surface.

Prevention

1. Thoroughly clean all areas to be painted. Be sure the surface is completely dry before applying undercoats or topcoats.

2. Drain and clean the air pressure regulator daily to remove trapped moisture and dirt. The air compressor tank should also be drained daily.
3. Use proper gun adjustments, techniques, and air pressure.
4. Select the thinner or reducer that is suitable for existing shop conditions.
5. Allow sufficient flash and dry times. Do not dry by fanning.

Solution

Sand the affected area down to a smooth finish and refinish.

PLASTIC FILLER BLEED-THROUGH

Condition

Discoloration (normally yellowing) of the topcoat color.

Causes

1. Too much hardener.
2. Applying topcoat before plastic filler is cured.

Prevention

1. Use correct amount of hardener.
2. Allow adequate cure time before refinishing.

Solution

1. Remove patch.
2. Cure topcoat, sand, and refinish.

PLASTIC FILLER NOT DRYING

Conditions

Plastic filler remains soft after applying.

Causes

1. Insufficient amount of hardener.
2. Hardener exposed to sunlight.

Prevention

1. Add recommended amount of hardener.
2. Be sure hardener is fresh and avoid exposure to sunlight.

Solution

Scrape off plastic filler and reapply.

RUNS OR SAGS

Condition

Heavy application of sprayed material that fails to adhere uniformly to the surface.

Causes

1. Too much thinner or reducer.
2. Wrong reducer.
3. Excessive film thickness without allowing proper dry time.
4. Low air pressure (causing lack of atomization), holding gun too close, or making too slow a gun pass.
5. Shop or surface too cold.

Prevention

1. Read and carefully follow the instructions on the label.
2. Select proper reducer.
3. Do not pile on finishes. Allow sufficient flash and dry time between coats.
4. Use proper gun adjustment, techniques, and air pressure.
5. Allow the vehicle surface to warm up to at least room temperature before attempting to refinish. Try to maintain an appropriate shop temperature for paint areas.

Solution

Allow the paint to dry. Wet sand with 600- or 1,000-grit sandpaper. Sand only on the run. Sand until the run is level with the surrounding paint. Repaint.

SAND SCRATCH SWELLING

Condition

Enlarged sand scratches caused by swelling action of topcoat solvents.

Causes

1. Improper surface cleaning or preparation. Use of too coarse a sandpaper or omitting a sealer in panel repairs greatly exaggerates swelling caused by solvent penetration.
2. Improper reducer, especially a slow-dry reducer when sealer has been omitted.
3. Underreduced or wrong reducer (too fast) used in surfacer causes "bridging" of scratches.

Prevention

1. Use appropriate grits of sanding materials for the topcoats being used.
2. Seal to eliminate sand scratch swelling. Select reducers suitable for existing shop conditions.
3. Use the proper reducer for the surfacer.

Solution

Sand the affected area down to a smooth surface and apply the appropriate sealer before refinishing.

SHOP TALK

Small surface defects like dirt or scratches may be sanded out of wet basecoat with 1,000-grit sandpaper and mild cleaner. Allow the basecoat to flash, wet the sand paper with mild cleaner instead of water, lightly sand, clean, and apply additional basecoat.

SOLVENT POPPING

Condition

Blisters on the paint surface caused by trapped solvents in the topcoats or surfacer—a situation that is further aggravated by force-drying or uneven heating.

Causes

1. Improper surface cleaning or preparation.
2. Wrong reducer. Use of a fast-dry reducer, especially when the material is sprayed too dry or at excessive pressure, can cause solvent popping by trapping air in the film.
3. Excessive film thickness. Insufficient drying time between coats and too heavy an application of the undercoats can trap solvents, causing popping of the color coat as they later escape.

Prevention

1. Thoroughly clean areas to be painted.
2. Select the reducer suitable for existing shop conditions.
3. Do not pile on undercoats or topcoats. Allow sufficient flash and dry time. Allow proper drying time for undercoats and topcoats. Allow each coat of surfacer to flash naturally—do not fan.

Solution

If damage is extensive and severe, paint must be removed down to undercoat or metal, depending on the depth of blisters; then refinish. In less severe cases, sand out, resurface, and re-topcoat.

UNDERCOAT SHOW-THROUGH

Condition

Variation in surface color.

Causes

1. Insufficient color coats.
2. Repeated compounding.

Prevention

1. Apply good coverage of color.
2. Avoid excessively compounding or polishing the surface.

Solution

Sand and refinish.

REVIEW QUESTIONS

1. Technician A says that blushing is found in cold shops. Technician B says that blushing happens in humid weather. Who is right?
   a. technician A
   b. technician B
   c. both technician A and technician B
   d. neither technician A nor technician B
2. Improper cleaning of the surface may cause _____ in the finish.
3. Technician A says that fisheyes are caused by silicone. Technician B says that contaminated air lines can cause fisheyes. Who is right?
   a. technician A
   b. technician B
   c. both technician A and technician B
   d. neither technician A nor technician B
4. Which problem is caused by solvent in the new topcoat attacking an old paint surface?
   a. blushing
   b. mottling
   c. lifting
   d. both a and b

5. Technician A says that orange peel can be caused by too little air pressure. Technician B says that cold shop temperatures cause orange peel. Who is right?
   a. technician A
   b. technician B
   c. both technician A and technician B
   d. neither technician A nor technician B

6. Solvent pop can be prevented by allowing more flash time.
   a. True
   b. False

7. Technician A says that washing a freshly painted vehicle in bright sunlight can cause water spots. Technician B says that sand scratch swelling can be prevented by using the proper grit of sandpaper. Who is right?
   a. technician A
   b. technician B
   c. both technician A and technician B
   d. neither technician A nor technician B

8. To correct peeling, extra coats of paint should be applied.
   a. True
   b. False

9. Technician A says that faded paint can be buffed. Technician B says that red paint fades faster than blue paint. Who is right?
   a. technician A
   b. technician B
   c. both technician A and technician B
   d. neither technician A nor technician B

10. An acid-spotted paint surface should be neutralized and buffed.
   a. True
   b. False

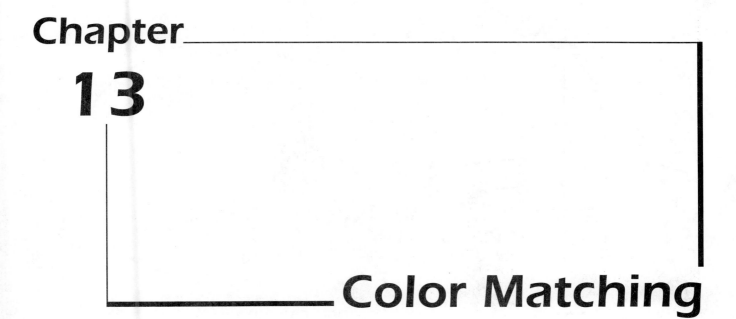

Chapter
13

Color Matching

Objectives

After reading this chapter, you will be able to:

- List the evaluation perspectives: Head-on, near-spec, and side-tone.
- Describe the elements of color: hue, value, chroma.
- List the steps in making a let-down panel.
- Explain how gun-handling techniques can change the color.
- Explain how to add tints to change the color.

Key Term List

standard
alternate
spectrophotometer
blend
mass tone
underhue

Color matching is probably the single most common problem in the automotive refinishing industry. Painters would be well advised to plan on blending every vehicle they paint. Blending can hide a minor mismatch. In some cases, however, the difference between refinish color and vehicle color is so great that a blend cannot hide it. In these cases, spray gun adjustments or spray technique adjustments may be utilized. In extreme examples, the refinish color may need to be tinted to match the vehicle.

 13.1 — COLOR ELEMENTS

Vehicle color should always be evaluated in sunlight or color-corrected light. The paint should be evaluated from these directions:

- **Head on.** Viewing the repaired area from an angle that is perpendicular to the vehicle.
- **Near-spec.** Viewing the repaired area from an angle just past the reflection of the light source.
- **Side-tone.** Viewing the repaired area at an angle of less than forty-five degrees.

Color is broken down into three elements:

- **Hue.** Basic color.
- **Value.** Lightness or darkness of color.
- **Chroma.** Intensity of color.

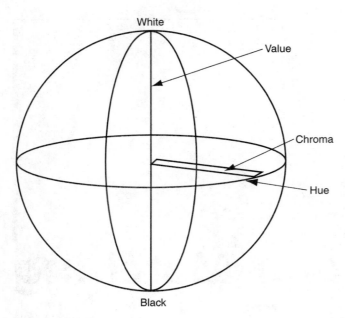

FIGURE 13–1 The color globe.

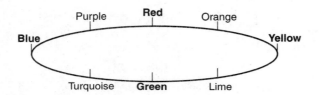

FIGURE 13–2 Hue. Primary colors are in bold; secondary colors are not.

Figure 13–1 shows a color globe with all of these elements combined.

HUE

The colors around the outside of the globe (Figure 13–2) are broken down into:

- **Primary colors.** Red, yellow, green, and blue.
- **Secondary colors.** Orange, lime, turquoise, and purple.

Notice that the secondary colors are mixtures of the two neighboring primary colors. The hue of a color can have a cast from either neighboring primary color. For example, a red can have a blue cast or a yellow cast. This is useful when comparing the refinish paint with the vehicle paint.

FIGURE 13–3 Value.

VALUE

In the globe, value ranges from the south pole with black through progressively lighter shades of gray to white at the north pole (Figure 13–3). Value indicates the lightness or darkness of a color. For example, a red can be described as light or dark, depending on value.

CHROMA

A straight line from the center of the globe out to the edge indicates chroma. Chroma is sometimes called saturation. An example in red is to think of a solid red color. If aluminum flakes are added to an intense red, the hue will not change, but the chroma will decrease. The metallic color is not as vibrant as the solid color (Figure 13–4).

Aluminum flakes or mica give paints an iridescent quality. The aluminum flakes, both large and small, act as mirrors and reflect light. If the flakes are near the surface, the hue of the color is hidden. The reflection of flakes gives a lighter color. Just the opposite is true if the flakes are not near the surface—the hue of the color is more apparent and the color is darker. Mica particles allow some light to pass through them. Mica-containing colors will have more chroma when compared to aluminum-flake colors, because the mica is more transparent, allowing the base color to show. Both aluminum-flake and mica colors should be evaluated from head-on, near-spec, and side-tone directions.

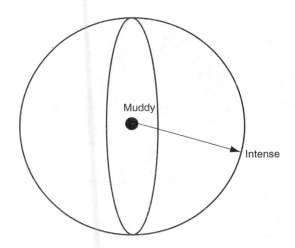

FIGURE 13–4 Chroma.

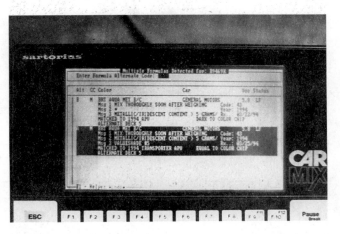

FIGURE 13–5 This color has several alternates. Note the heading "Multiple Formulas Detected."

One last topic in color evaluation is metamerism. Metamerism occurs when two colors look the same under one type of light but different under another type of light. An example would be if the repainted vehicle looked good in the fluorescent lights in the shop but was a mismatch outside in the daylight. To avoid metamerism when tinting, only add tints from the formula.

13.2 ALTERNATES

It takes considerable practice to describe and compare colors. Refinish paint manufacturers have recognized that the **standard**—the color chip in the paint book does not always match the OEM finish. To help painters to match colors, some paint manufacturers formulate **alternates** for OEM colors. An alternate color means that an OEM color does not match the standard. The paint manufacturer matches the OEM paint and designates this as an alternate. Some colors may have five to ten alternates (Figure 13–5). Quite often the alternates are described in relation to the standard. An example would be redder than the color chip or darker than the prime (standard). Alternate decks are available (Figure 13–6). These are paint chips of each alternate. The alternate chips are compared to the vehicle and the closest match is used. If no alternate deck is available, the color chip is compared to the vehicle and the alternate description that best fits the situation is used.

FIGURE 13–6 This alternate deck consists of paint samples to be compared to the vehicle. Each paint sample has a separate formula.

13.3 TEST PANELS

While a test panel sprayout is recommended in many refinish applications, it is vital with three-coat finishes. Test panels for tricoats are needed to determine the correct amount of midcoat pearl that must be applied to achieve an exact color match. The midcoat color is the most critical portion of the tricoat repair. Because gun pressure, reduction, and spray technique can affect the amount of color being applied to a given job,

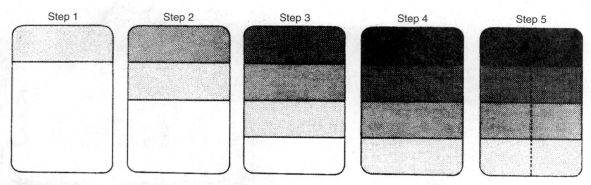

| Step 1 | Step 2 | Step 3 | Step 4 | Step 5 |

FIGURE 13–7   *Test panel for tricoats. The same idea can be used for basecoat/clear coat.*

the extra time spent spraying one or more test panels will be repaid many times over with a finished job that satisfies the customer and does not come back later.

Here is how to make a test panel for a tricoat finish:

1. Prepare a test panel with the same color undercoat being used on the job. If a sealer is going to be used, apply the sealer to the test panel also. Generally, a light color undercoat (or sealer) is preferred for tricoat repairs.

2. Apply the basecoat color to hiding, using the same pressure and spray pattern that you will on the job. Duplicating the actual spray techniques when preparing the test panel is an important point. Make sure not to vary the procedures simply because the work is a small panel and not a full repair.

3. After the panel has dried, divide it into four equal sections (Figure 13–7). Next, mask off the lower three-quarters of the panel, exposing the top quarter.

4. Apply one coat of mica midcoat color over the top quarter of the panel.

5. After the first mica coat has flashed, remove the masking paper and move it down to the middle of the panel, exposing the top half.

6. Apply another coat of mica midcoat color over the exposed top half of the panel.

7. After this second coat has flashed, remove the masking paper and move it down to expose three-quarters of the panel.

8. Apply another coat of mica midcoat color over the exposed three-quarters of the panel.

9. After flashing, remove the masking paper entirely.

10. Apply a fourth coat of mica midcoat color, as always, spraying the coating in the same way as would be done on the repair.

11. After the entire panel has dried, mask off the panel *lengthwise* this time.

12. Apply the manufacturer's recommended number of coats of clear to the exposed side.

Once the test panel is completed, lay it on the vehicle to determine the number of mica midcoat coatings needed to achieve a precise color match. Be sure to view the match from different angles and under sunlight if at all possible. And be sure that the panel on the vehicle is thoroughly cleaned before making the comparison.

If the refinisher prefers, instead of using one panel divided into small sections, separate panels can be prepared to provide a larger work area. If separate panels are used, start by spraying all four panels with the midcoat color. Remove one of the panels and then spray the remaining three with a second coat. Spray a third coat on the last two panels and a fourth coat on the last panel only. Then the panels should be masked vertically and receive the recommended coats of clear.

If, after completing the test panel procedure, a proper color match is still not achieved, recheck the coating with the manufacturer's product number. Be sure that the correct basecoat color is used for this match. Slight variations in the basecoat color can produce an unmatchable final finish. Also, be sure that the paint manufacturer's label directions for each product are followed.

Do not mix brands of products. From primer to final clear coat, stay with one manufacturer's system. The

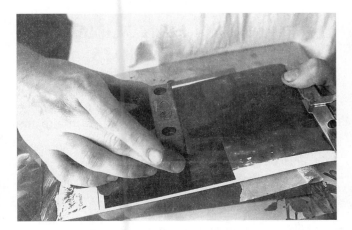

FIGURE 13–8 Draw-down bar in use.

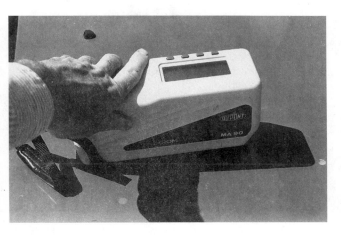

FIGURE 13–9 Color analyzer in use.

use of a single system also includes solvents and reducers. Paint company laboratories match colors with a balance of solvents or reducers recommended for their products. Using another manufacturer's reducer might save a few dollars, but it will often result in hours of color-matching problems. Keep in mind that individual coating manufacturers might have slightly different approaches, and these should be followed carefully.

DRAW-DOWN BAR

The draw-down bar is a precision tool with a machined blade to give an even paint film distribution (Figure 13–8). To use it, a black-and-white test panel is taped to a perfectly flat surface, such as an aluminum clipboard. Paint is distributed onto the test panel, and the draw-down bar is drawn through the paint, spreading it into a uniform paint thickness. After the first draw-down flashes, the process is repeated until the black and white on the test panel is no longer visible through the paint film. Then the test panel is compared to the car.

When working on base/clear, make the draw-down over a strip of clear film (place over a check hiding panel so the painter knows coating is achieving the desired hiding effect). When you turn the clear film over, the strip will have a base/clear appearance.

If additional tinting is needed, another draw-down should be made on another test panel. This allows the refinisher to see the direction in which the tint is moving by comparing the panels to each other. When the desired color is achieved, the material can be sprayed and further adjusted through gun techniques. The draw-down bar eliminates unnecessary waste of time, material, and labor to adjust color.

13.4 COLOR ANALYSIS COMPUTER

A color analysis computer (Figure 13–9) saves considerable time and effort when refinishing vehicles. The **spectrophotometer** is calibrated, then placed on a cleaned area of the vehicle or part. Several readings are taken. The readings are then compared with the stored paint formulas. The closest formula is selected by the computer. If a usable formula is not stored, the computer is capable of producing a unique formula. These systems have about a 75 percent success rate in choosing a blendable color match formula.

13.5 SPRAY GUN ADJUSTMENTS

The spray gun can be adjusted to make changes in the value. The effect is more pronounced in metallic colors. In general, putting the paint on wetter will darken the color. If the paint is sprayed wet, the metallic flakes have time to settle to the bottom of the wet-paint film. When the paint dries, the hue of the color is more prominent than the metallic. The color appears darker. Putting the paint on drier will lighten the color (Figure 13–10). In dry spray, the metallic flakes do not have a chance to settle into the paint film; they are trapped at the surface. This gives a more pronounced metallic effect. The hue of the color is not as noticeable.

These spray gun adjustments will make the paint spray wetter:

- Decrease air pressure
- Decrease fan size
- Increase fluid

These spray gun adjustments will make the paint spray drier:

- Increase air pressure
- Increase fan size
- Decrease fluid

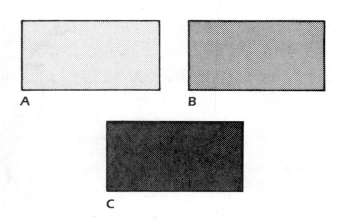

FIGURE 13–10 *Changing value with a spray gun: (A) Dry spray—lighter color. (B) Normal spray. (C) Wet spray—darker color.*

13.6 GUN-HANDLING ADJUSTMENTS

The same logic applies to spray gun handling; wetter paint is darker, drier paint is lighter. The following are spray gun-handling techniques that will make the paint spray wetter:

- Decrease gun movement speed
- Decrease gun to surface distance
- More overlap
- Less flash time between coats

The paint spray may be made drier by these gun-handling techniques:

- Increase gun movement speed
- Increase gun to surface distance
- Less overlap
- More flash time between coats

Other factors controlling lightness or darkness are listed in Table 13–1.

13.7 TINTING

If the test panel comparison to the vehicle shows a difference in hue or chroma that cannot be solved by blending, the color must be tinted. Tinting should al-

---

TABLE 13–1:   ADJUSTING LIGHTNESS/DARKNESS OF COLOR

Variable	To Make Colors	
	Lighter	**Darker**
Shop Condition 1. Temperature 2. Humidity 3. Ventilation	1. Increase 2. Decrease 3. Increase	1. Decrease 2. Increase 3. Decrease
Solvent Usage 1. Type of solvent 2. Reduction of color 3. Use of retarder	1. Use faster-evaporating solvent. 2. Increase amount of solvent. 3. Do not use retarder.	1. Use slower-evaporating solvent. 2. Decrease amount of solvent. 3. Add retarder to solvent.

ways be the last resort, after alternates and test panels. To tint color, have the following available:

1. Vehicle in the sunlight and buffed to remove fade. Check the inside of the doors to see if the paint inside the door is the same as the outside.
2. Test panel
3. Paint formula
4. Tint chart (see color section of this book)

The tint chart shows the **mass tone** of each tint. However, tint has hue other than just the mass tone. To determine the **underhue,** the chart has white and aluminum letdowns for each tint. The white letdown shows what the underhue is in a solid color when the tint is mixed, one-to-one with white. The aluminum letdown, tint mixed one-to-one with aluminum, shows the underhue in a metallic color. For example, phthalo blue and permanent blue have similar mass tone. But when they are mixed one-to-one with white, permanent blue has a red cast and phthalo blue is lighter. Knowledge of the underhue of a tint will enable the painter to more accurately move the color in the desired direction.

The first step in tinting is to compare the test panel with the vehicle. Evaluate head-on and side-tone. Complete this statement: The vehicle is _____ compared with the paint. For simplicity, evaluate only one aspect of the color at a time. Check value first. Is the vehicle lighter or darker than the paint? Next, look at the hue. If the vehicle is red, is it redder, bluer, or yellower than the paint? Look at the chroma last. Is the vehicle more or less intense in color than the paint? Which aspect is off the most? Adjust the aspect that is furthest off first. This may change the other aspects of the color as tints are added.

A chart may be helpful in making adjustments. Plot the paint by looking at the tints in the formula (Figure 13–11). Place a *P* on the chart to indicate the paint location. In this case, the paint is blue with a red underhue (Figure 13–12). Next plot the vehicle. In this example, the vehicle is blue with a green underhue (Figure 13–13). The paint hue needs to move toward green. In looking at the formula, a blue with a green underhue is found. A small amount of paint is placed on a paint can lid. A small amount of tint is added to the paint. They are mixed together (Figure 13–14). This will give a quick reference to see if the color is moving in the right direction. If this is correct, a measured and recorded amount of tint is added to the paint. A test panel is sprayed and hue and chroma are reevaluated.

To avoid problems with metamerism, add only tints found on the formula. Following are some general rules to use when tinting.

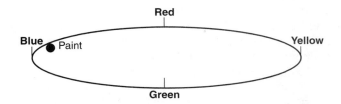

TINTING GUIDE	MIX SIZE: 2 QUART		
806J HS BLACK	164.8	150K B/C BALANCER	1383.2
815J MULTIGRAD ALUM	246.8	175K BINDER	1736.4
862J TRANSP RED	296.0		
829J LIGHT BLUE	338.0		
813J MED COARSE AL	377.2		
802J LS WHITE	404.0		
816J MED FINE ALUM	422.0	VOC-LE: 5.7 VOC-AP: 4.8	

FIGURE 13–11   Paint formula for blue color.

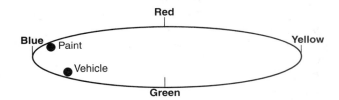

FIGURE 13–12   Plot of the paint color.

FIGURE 13–13   Plot of the paint and vehicle colors.

FIGURE 13–14   Mixing a small amount of paint and tint to see if the color moves in the right direction.

HUE

Each primary color can have the underhue of a neighboring primary color. For example, a red can have a blue or yellow underhue; a blue can have a red or green underhue; a yellow can have a red or green underhue, and a green can have a blue or yellow underhue. To adjust the hue, add a tint from the formula that achieves the desired hue, or at least the underhue. This works well with a primary color. But what about colors that are not primary, like brown or charcoal gray? These colors can be plotted also. A brown can be bluer, yellower, or redder when compared to another brown. A charcoal gray is plotted near the center of the globe. Check the formula to see what color tints are combined to make it.

Sometimes kill charts are used to remove a hue from a color. The theory is that if you add a tint from the opposite side of the color globe, the hue will be removed. Table 13–2 shows a kill chart. In the first example, the paint is blue, but it has an undesirable red cast. If green is added, it will kill or eliminate the red. Even though this does work, the chroma is reduced, muddying the blue. If the tint added is not from the formula it may lead to a metamerism problem. If a change in hue and a decrease in chroma are required, a kill chart will work.

VALUE

In solid colors, the value can be lightened by adding white, if white is in the formula. However, if white is not in the formula, add the lightest tint in the formula. Darkening the value can be accomplished by adding black, if it is in the formula. If black is not in the formula, add the darkest tint in the formula. This may change the hue. In metallics, to lighten the color, add a bright or coarse aluminum. To darken, add a dull or fine aluminum.

SHOP TALK

To a degree, depending on how well a color covers or hides the undercoat, the color of the topcoat is influenced by the color of the undercoat. For example a red pearl topcoat, a poor hider, may appear lighter when sprayed over a gray undercoat, compared to the same color sprayed over a dark red undercoat.

CHROMA

To increase chroma, add more of the dominant tint. For a blue, it would be the purest blue in the formula. The purest blue would be the blue with the least amount of red or green underhue. To decrease chroma, add gray—black and white—or in metallics, aluminum.

TABLE 13–2:   KILL CHART METHOD OF CHANGING CASTS

Color	Add		Cast
Blue	Green	to kill	Red
Blue	Red	to kill	Green
Green	Yellow	to kill	Blue
Green	Blue	to kill	Yellow
Red	Yellow	to kill	Blue
Red	Blue	to kill	Yellow
Gold	Yellow	to kill	Red
Gold	Red	to kill	Yellow
Maroon	Yellow	to kill	Blue
Maroon	Blue	to kill	Yellow
Bronze	Yellow	to kill	Red
Bronze	Red	to kill	Yellow
Orange	Yellow	to kill	Red
Orange	Red	to kill	Yellow
Yellow	Green	to kill	Red
Yellow	Red	to kill	Green
White	White	to kill	Blue
White	White	to kill	Yellow
Beige	Green	to kill	Red
Beige	Red	to kill	Green
Purple	Blue	to kill	Red
Purple	Red	to kill	Blue
Aqua	Blue	to kill	Green
Aqua	Green	to kill	Blue

FLIP-FLOP OF COLOR

Flip-flop is a condition that occurs in metallics. It involves the positioning of the aluminum particles and the manner in which light is reflected to the observer (Figure 13–15). This effect results from the percentage of aluminum particles that are oriented in a specific direction and their depth in the paint film. The direction and intensity of the light being reflected back through the paint film creates the flip-flop phenomenon.

The first approach to correcting the problem is to adjust your spraying technique to compensate for this effect. Spraying the fender a little wetter will slightly darken the appearance when looking directly into the panel. When viewed from an angle, the panel will appear lighter. This occurs because the aluminum particles are positioned flatter and deeper in the paint film.

Spraying the panel slightly drier reverses the effect, giving a light appearance when looking directly at the panel. This is because the aluminum particles are closer to the surface. The panel will appear darker when viewed at an angle, as light becomes trapped. Both of these techniques are a compromise and should be used to correct minor conditions of flip-flop, because the match in one direction can be changed too severely to be acceptable.

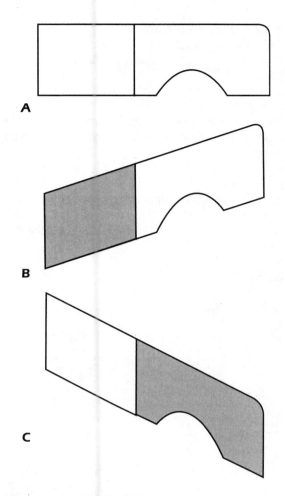

FIGURE 13–15 Flip-flop. (A) Head-on: door and fender match. (B) Rear side-tone: door is darker than the fender. (C) Front side-tone: fender is darker than the door.

If spray techniques cannot correct this condition, the addition of a small amount of white will eliminate the sharp contrast from light to dark when the surface is viewed at various angles. The white acts to dull the transparency, giving a more uniform, subdued reflection through the paint film. Care should be taken when adding white, since the change occurs quickly. Once too much white is added, recovering the color match becomes virtually impossible.

When confronted with an extremely difficult flip-flop condition, the best method involves adding white or titanium dioxide toner and blending the color into the adjacent panels. When blending, extend the color in stages.

A good painter must know how to handle metallic colors. They are very sensitive to the solvents with which they are reduced and the air pressure with which they are applied. Metallic colors are also affected by a number of variables. A *variable* is a part

of the spray-painting conditions, such as temperature, humidity, and ventilation, or a part of the spray-painting process, such as amount of reduction, evaporation, speed of solvents, air pressure, and type of equipment. To get good color matches, one must understand how certain paint variables affect the shades of metallic colors.

Variables are divided into two categories: positive and negative. Positive variables are those things that a painter does to duplicate the original finish, which in turn results in a good color match. They are:

- Slowness of solvent evaporation. This allows the painter to reproduce the factory finish.
- Wetness of color application.
- Proper spraying technique and the correct air pressure.

Negative variables are those that cause the shades of colors to be off standard. Most common are:

- Improper reduction.
- Improper agitation.
- Improper application—primarily too-high or too-low air pressure.

In summary, the shades of metallic colors are controlled by:

- Choice of solvents.
- Color reduction.
- Air pressure.
- Wetness of application.
- Spraying techniques.

REVIEW QUESTIONS

1. Technician A says that hue is the lightness of a color. Technician B says that chroma is the basic color. Who is right?
   a. technician A
   b. technician B
   c. both technician A and technician B
   d. neither technician A nor technician B
2. The primary colors are
   a. red, black, gray, and white.
   b. red, blue, green, and yellow.
   c. red, blue, black, and white.
   d. orange, lime, turquoise, and purple.
3. Technician A says that minor mismatch can be hidden by blending. Technician B says that saturation is also known as chroma. Who is right?
   a. technician A
   b. technician B
   c. both technician A and technician B
   d. neither technician A nor technician B

4. A blue can be either too red or too yellow.
   a. True
   b. False
5. Technician A says that mica colors have less chroma than metallic colors. Technician B says that if the metallic flakes are near the surface, the color is lighter. Who is right?
   a. technician A
   b. technician B
   c. both technician A and technician B
   d. neither technician A nor technician B
6. _____ means that paint looks different colors under different types of light.
7. Technician A says that wet coats will darken a metallic color. Technician B says that dry coats will lighten a metallic color. Who is right?
   a. technician A
   b. technician B

   c. both technician A and technician B
   d. neither technician A nor technician B
8. An increase in air pressure will darken a metallic color.
   a. True
   b. False
9. Technician A looks at chroma first when matching a color. Technician B looks at value first. Who is right?
   a. technician A
   b. technician B
   c. both technician A and technician B
   d. neither technician A nor technician B
10. A metallic color should be lightened by adding white.
   a. True
   b. False

Chapter

14

Painting Plastic Parts

Objectives

After reading this chapter, you will be able to:
- **Prepare plastic parts for refinishing.**
- **Describe the paint finishing systems applicable to plastic parts.**
- **Identify plastic parts.**
- **Repair minor damage to plastic parts.**

Key Term List

thermoplastic
flex agent
thermoset

In recent years, more and more plastic has been used in various parts of car bodies, particularly in the front end: in bumper and fender extensions, and in soft front fascia, fenders, hoods, doors, quarter panels, roofs, aprons, grille opening panels, stone shields, instrument panels, and ground effects (Figure 14–1). Because these parts are much lighter in weight than sheet metal, they have become an important part of every American manufacturer's fuel-saving weight reduction program. And because of the high strength-to-weight ratio of plastic, the weight decrease does not mean a decrease in strength. Every indication is that plastic body parts are here to stay, and new applications for plastic will probably be found in the future. Therefore, automotive painters can expect to be painting a greater number of plastic parts.

14.1 — TYPES OF PLASTICS

There are two types of plastics used today in automotive production:

- **Thermoplastics.** These plastics are capable of being repeatedly softened and hardened by heating or cooking. They soften or melt when heat is applied and, therefore, are weldable.
- **Thermosetting plastics.** These plastics are materials that undergo a chemical change by the action of either heat, a catalyst, or ultraviolet light, leading to an infusible state. Catalyst and resin mix to form a new product. Thermosets are not weldable.

PLASTIC IDENTIFICATION

Before deciding upon the proper repair technique to use, it is first necessary to identify the type of plastic from which the component is made. This is very

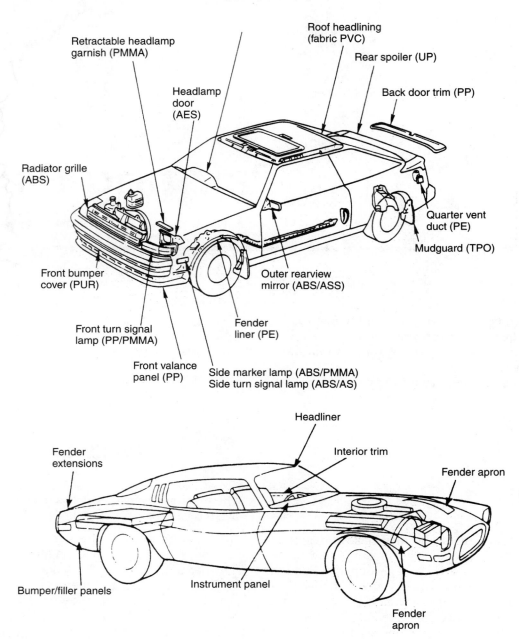

FIGURE 14–1 *Some examples of plastic types and locations.*

important, because a repair job that is based on incorrect identification of the plastic can quickly delaminate, crack, or discolor. There are several ways to identify an unknown plastic. For instance, various types of plastic can be identified according to national identification symbols, which can often be found on the parts to be repaired. When parts are not identified by these symbols, refer to the manufacturer's technical literature for plastic identification. Domestic manufacturers are using identification symbols more and more; unfortunately, there are many who do not. Another problem with this system is that it is usually necessary to remove the part to read the letters.

One of the best ways to identify an unknown plastic is to become familiar with the different types of plastics and where they are commonly used. Table 14–1 gives the identification symbol, chemical and common names, and applications of the more common

TABLE 14–1: STANDARD SYMBOL, CHEMICAL NAME, TRADE NAME, AND DESIGN APPLICATIONS OF MOST COMMONLY USED PLASTICS

Symbol	Chemical Name	Common Name	Design Applications	Thermosetting or Thermoplastic	Flexible or Rigid
ABS	Acrylonitrile-butadiene-styrene	ABS, Cycolac, Abson, Kralastic, Lustran, Absafil, Dylel	Body panels, dash panels, grilles, headlamp door	Thermoplastic	R
ABS/PC	ABS/Polycarbonate	—	Doorskins and rocker panels	Thermoplastic	F
EP	Epoxy	Epon, SPO, Epotuf, Araldite	Fiberglass body panels	Thermosetting	R
EPDM	Ethylene-propylene-diene-monomer	EPDM, Nordel	Bumper impact strips, body panels	Thermosetting	F
FRP	—	Fiberglass-reinforced plastic	Header panels	Thermosetting	R
PA	Polyamide	Nylon, Capron, Zytel, Rilsan, Minlon, Vydyne	Exterior finish trim panels	Thermosetting	R
PC	Polycarbonate	Lexan, Merlon	Grilles, instrument panels, bumpers	Thermoplastic	R
PE	Polyethylene	Dylan, Fortiflex, Marlex, Alathon, Hi-fax, Hosalen, Paxon	Inner fender panels, interior trim panels, valances, spoilers	Thermoplastic	F
PP	Polypropylene	Profax, Olefo, Marlex, Olemer, Aydel, Dypro	Interior mouldings, interior trim panels, inner fenders, radiator shrouds, dash panels, bumper covers	Thermoplastic	F
PPO	Polyphenylene oxide	Noryl, Olefo	Chromed plastic parts, grilles, headlamp doors, bezels, ornaments	Thermosetting	R
PPO/Nylon	Polyphenylene oxide/Polyamide	GTX	Fenders, quarter panels	Thermosetting	F
PS	Polystyrene	Lustrex, Dylene, Styron, Fostacryl, Duraton	Door panels	Thermoplastic	R
PUR	Polyurethane	Castethane, Bayflex	Bumper covers, front and rear body panels, filler panels	Thermosetting	F
PVC	Polyvinyl chloride	Geon, Vinlyete, Pliovic	Interior trim, soft filler panels	Thermoplastic	F
RIM	"Reaction injection molded" polyurethane	RIM, Bayflex	Bumper covers	Thermosetting	F
R RIM	Reinforced RIM-polyurethane	RRIM	Exterior body panels	Thermosetting	F
TPO	Thermoplastic olefin	Ferroflex, Polytrope	Bumper covers, valance panels	Thermoplastic	F
TPUR	Polyurethane	Pellethane, Estane, Roylar, Texin	Bumper covers, gravel deflectors, filler panels, soft bezels	Thermoplastic	F
UP	Polyester	SMC, Premi-glas, Selection Vibrin-mat	Fiberglass body panels	Thermosetting	R

automotive plastics. Plastic application information can often be found in shop manuals or in special manufacturer's guides.

PLASTIC PARTS PREPARATION

Plastic parts are usually considered either hard (rigid) or flexible (semirigid). Some flexible plastic auto body replacement parts come from the factory already primed, while others are delivered unprimed. If the parts are factory-primed, no additional priming may be necessary. If they are not, both rigid and semirigid plastics might benefit from the use of a special plastic primer to improve paint adhesion. TPO, in particular, has an extremely slick, waxy surface that makes it difficult for the topcoat to form a strong bond to the substrate unless a primer is used. Adhesion to ABS used in exterior applications also is greatly enhanced by priming. The special plastic primer, sometimes called polypropylene primer, is packaged ready to spray. After the part is cleaned and scuff sanded, one coat of the plastic primer is sprayed on.

FLEXIBLE PAINT ADDITIVES

14.2

A flexible plastic part, such as a bumper fascia, can have slight pressure applied to it, denting the plastic. However, when the pressure is released, the plastic will return to its normal shape with no damage (Figure 14–2). The plastic and the paint on the plastic are both flexible.

Some paint manufacturers recommend that a **flex agent** be mixed in with the paint when one is refinishing flexible parts (Figure 14–3). The flex agent will make the paint pliable after it dries. If the repainted part is damaged, the paint will not crack. Table 14–2 lists the various types of plastic and if a flex agent or special primer is needed. There are two types of paint that need the flex agent. The first is a urethane surfacer. If a bumper fascia or other flexible part is repaired and filling is required, the surfacer is mixed as usual and a flex agent is added to it. The surfacer dries in the normal fashion and is blocked out like a steel part. The other type of paint that may need flex agent is top coat. If the top coat is single stage, the enamel, reducer, hardener, and flex agent are mixed together. If the top coat is basecoat/clear coat, the flex agent is not used in the basecoat. Only the clear coat needs flex agent. Check with the refinish manual to see if a flex agent is required for a surfacer or top coat. In some cases, it may be required on more paints than just the surfacer and clear coat.

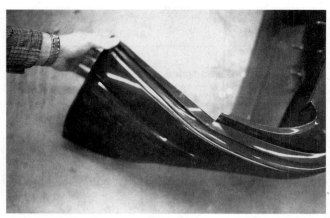

A

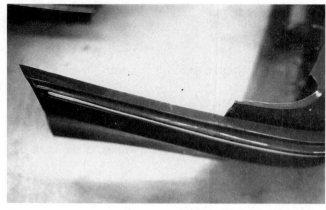

B

FIGURE 14–2 *Flexible plastic bumper fascia: (A) Bent. (B) Returns to normal shape after pressure is released.*

Flexible plastic parts are made by molding. The liquid plastic is poured or injected into a mold. In some cases, a chemical reaction takes place inside the mold. In other cases, the mold is heated to harden the plastic. A mold release agent is added to the liquid plastic to allow the formed part to be removed from the mold. Some types of plastic have a dye added to them in order to form a part with the desired color. Other plastic parts must be painted to the desired color. The painting of parts is a common practice in a collision repair shop.

SHOP TALK

Refinish paint peeling from replacement bumpers has been a common collision repair industry problem. To prevent this problem, always properly clean, sand, and seal replacement bumpers. A continuing problem should be referred to the paint manufacturers' representative.

FIGURE 14–3   Flex additive.

TABLE 14–2:	FLEX AGENT AND SPECIAL PRIMER FOR AUTOMOTIVE PLASTICS	
Plastic Name	**Flex Agent Required**	**Special Primer Required on Bare Parts**
ABS	No	Yes
ABS/PC	Yes	No
EP	No	No
EPDM	Yes	No
FRP	No	No
PA	No	No
PC	No	No
PPO	No	No
PE	Yes	No
PP	Yes	No
PPO/Nylon	Yes	No
PS	No	No
PUR	Yes	Yes
TPO	Yes	Yes
TPUR	Yes	Yes
PVC	Yes	No
RIM	Yes	No
RRIM	Yes	No
UP	No	No

FIGURE 14–4   Cleaning a replacement bumper.

14.3   PAINTING NEW PLASTIC PARTS

FLEXIBLE PLASTIC

To paint a new flexible plastic part, first check to see if the part is the right type and is undamaged. Test-fit the bumper reinforcement and impact absorber. If you have difficulty putting them together, consider assembling the bumper before painting. Wrestling a painted bumper together leads to scratches. Next, check the painting instructions supplied with the part or the paint system manual. If there aren't any specific instructions, follow these general steps:

Most flexible plastic parts, OEM and aftermarket, are primed before shipment. If the part has a shiny, slick surface, it may not be primed. In this case, clean off the mold release agent with a one-to-one mixture of rubbing alcohol and water. A plastic cleaner may also be used. Wet the part with alcohol and water and wipe off. Sand with 400-grit sandpaper and the alcohol/water solution. Reclean with alcohol and water. Spray the appropriate sealer or special plastic primer on the part immediately after sanding. The entire process should not take more than thirty minutes.

If the part is primed, it will have a dull and rough surface, it should be wiped with a mild wax and grease remover (Figure 14–4). Sand the part with 400-grit paper or a red abrasive pad (Figure 14–5). Do not sand the primer off. If bare plastic is exposed, clean with alcohol and water to remove the mold release agent. Spray the part with the appropriate sealer (Figure 14–6).

After the sealer or primer has flashed, the basecoat and clear coat can be sprayed. Usually if the flexible part is sprayed at the same time as the vehicle, it is

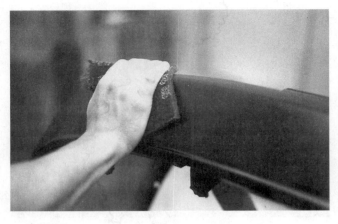

FIGURE 14–5 *Scuff-sanding replacement bumper with an abrasive pad.*

FIGURE 14–6 *Spraying sealer on a prepped replacement bumper.*

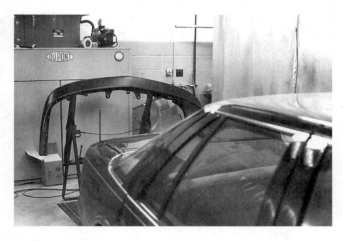

FIGURE 14–7 *This bumper will be sprayed at the same time as the car. Because it is separate, flex agent can be added to the clear.*

sprayed separately (Figure 14–7). This will allow a flexed clear to be applied to the flexible part but not to the vehicle. Also, it allows the underside of the bumper to be easily sprayed. Often the part is two-tone—a portion of the part is the color of the vehicle and a strip around the perimeter is flat black. In this case, spray the entire bumper with the vehicle color and clear. After curing, the strip is masked, scuffed, cleaned, and sprayed with a flexible flat black paint.

SHOP TALK

Mold release agents may be exuded when a plastic part is baked. To prevent problems once the part is painted and baked, the part should be prebaked. The procedure would be to wipe the part with plastic cleaner, prebake, rewipe with plastic cleaner, scuff sand, and prime.

RIGID PLASTIC

A new rigid part such as an SMC or FRP part can be painted in the same way as a steel part. Usually the part has a factory-applied primer. Clean the part to remove wax and grease. Sand the part with 400-grit sandpaper. Be sure to sand the primer to a uniform dullness. Spray a coat of sealer. When the sealer has flashed, the basecoat and clear coat can be applied. No flex agent is required. If the part is not primed, clean with a plastic cleaner or a one-to-one mixture of alcohol and water. Spray with an adhesion promoter or epoxy primer, then topcoat.

SHOP TALK

Paint manufacturers may produce only a lacquer paint for some plastic painting applications. For example, the trim color for Ford polycarbonate bumpers, when a different color from the vehicle, may be available only as lacquer. Paint for a vinyl roof is another example of a trim color that may be available in lacquer only.

14.4 **REPAIRING PLASTIC PARTS**

REPAIRED FLEXIBLE PLASTIC

Flexible parts are often damaged in a collision. Scratch damage can be repaired with a flexible material. This repair material remains flexible after curing.

FIGURE 14–8   Featheredging a damaged bumper.

FIGURE 14–10   Block-sanded bumper ready for repaint.

FIGURE 14–9   An overfill of plastic repair material is applied to the scratch.

FIGURE 14–11   This FRP header panel has cracks above and below the parking light.

This will allow the repaired part to give under slight pressure the same way an undamaged part would. The only difficulty in repair is that some types of plastic do not allow the repair material to stick. In these types of plastic, an adhesion promoter is used to help the material stick to the plastic.

The first step in the repair is to thoroughly clean the part. Use soap and water followed by an aggressive wax and grease remover. A plastic cleaner may also be used.

Featheredge the scratch in the same manner as steel is prepped: 80-grit sandpaper on a sanding block followed by 180- and 320-grit sandpaper (Figure 14–8). If the plastic, not the paint, sands dry, no adhesion promoter is needed. If the plastic smears instead of sands, as in the case with PP, TPO, or EPDM plastic, an adhesion promoter must be sprayed on the plastic before the repair material is applied. The adhesion promoter will allow the repair material to bind with the plastic.

The repair material is mixed one-to-one. Apply an overfill to the scratch fill (Figure 14–9). Allow the material to cure. Block-sand with 80-, 180-, and 320-grit sandpaper. Clean the part with a mild cleaner or plastic cleaner. Apply masking as needed. Spray two to three coats of flexible urethane surfacer. Guide coat as needed. After proper cure time, block-sand the surfacer (Figure 14–10).

REPAIRED RIGID PLASTIC

The repair of a scratch in a rigid plastic part such as FRP or SMC is similar to the repair of steel (Figure 14–11). The part needs to be cleaned with soap and water, then with aggressive cleaner. The scratch is featheredged with 80-, 180-, and 320-grit sandpaper. Polyester filler used on steel in some cases may be used on FRP and SMC. Check the label on the can of filler to see if it can be used on these rigid plastics. A rigid plastic filler is a better choice. Apply the filler to

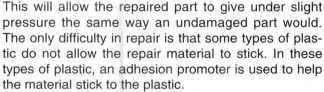

FIGURE 14–12 Overfill of rigid repair material to the crack below the parking light.

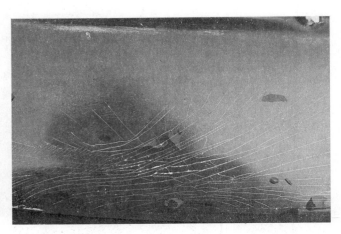

FIGURE 14–14 This flexible bumper was damaged and the paint cracked instead of flexed.

FIGURE 14–13 Block-sanding the filler.

FIGURE 14–15 This bumper is being painted separately.

the scratch (Figure 14–12). Allow the filler to cure. Block-sand the filler with 80-, 180-, and 320-grit sandpapers (Figure 14–13). Apply the needed masking. Spray two to three coats of urethane surfacer, then guide coat.

14.5 REPAINTING PLASTIC PARTS

REPAINTING FLEXIBLE PLASTIC PARTS

Inspect the flexible part prior to surface prep. The paint on a bumper that has been repainted and then damaged again will often crack. This can also happen to OEM paint if the part was painted many times at

the factory (Figure 14–14). All of the cracked paint must be sanded off. First clean the bumper with soap and water followed by an aggressive cleaner. Sand off any cracked paint with 320-grit sandpaper. Clean, mask, and apply a surfacer as needed. Block-sand the surfacer with 320-grit sandpaper followed by 600-grit wet sandpaper. If the entire bumper is to be basecoated, sand thoroughly with 600-grit sandpaper. If the bumper is too blended, only a portion needs to be basecoated and the entire bumper clear coated. Sand the area to be basecoated with 600-grit sandpaper, and sand the area to be clear coated with 1,000-grit sandpaper. In certain circumstances, when only a section of the bumper needs to be painted and there is a narrow area dividing what is to be painted from what is not, a melt is acceptable. After the surface prep is completed, clean the bumper and mask for painting. If the bumper is too low to be painted on the car, remove it so it can be painted thoroughly (Figure 14–15). Spray three coats of base. If there are convo-

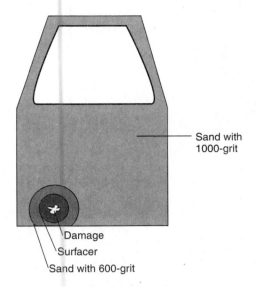

Sand with
1000-grit

Damage
Surfacer
Sand with 600-grit

FIGURE 14–16 _Repair and repaint of a minivan hatch._

lutions, spray these difficult areas first to make sure that they are covered. In exceptionally difficult areas, adjust the spray gun to a narrow fan. Decrease the fluid and turn the air pressure down to twenty-five pounds to allow the paint to be fogged on. If flex agent is required, add it to the clear coat. Spray two coats of clear.

REPAINTING RIGID PLASTIC PARTS

Painting a rigid plastic part is similar to painting steel. For example, a damaged SMC hatch on an Aerostar van can be repaired (Figure 14–16). In this case, clean the panel and sand around the area that will be basecoated with 600-grit sandpaper. Sand the area that will be clear coated with 1,000-grit sandpaper. Spray three coats of base, extending each coat out two inches from the previous coat. Spray two coats of clear over the entire panel.

REVIEW QUESTIONS

1. Technician A says that ABS is a thermoplastic. Technician B says that ABS is rigid. Who is right?
   a. technician A
   b. technician B
   c. both technician A and technician B
   d. neither technician A nor technician B
2. EP means extended plastic.
   a. True
   b. False
3. Technician A says that most plastic replacement parts are primed at the factory. Technician B says that factory primer must be scuffed before painting. Who is right?
   a: technician A
   b: technician B
   c: both technician A and technician B
   d: neither technician A nor technician B
4. Painter A does not apply a primer to rigid plastic parts. Painter B does. Who is right?
   a. painter A
   b. painter B
   c. both painter A and painter B
   d. neither painter A nor painter B
5. Technician A says that plastic can be flexible or rigid. Technician B says that mold release agent must be removed before painting. Who is right?
   a. technician A
   b. technician B
   c. both technician A and technician B
   d. neither technician A nor technician B
6. Mold release agent can be taken off prior to painting by a one-to-one mixture of _____ and water.
7. Technician A assembles a flexible bumper prior to painting. Technician B says that this is unnecessary. Who is right?
   a. technician A
   b. technician B
   c. both technician A and technician B
   d. neither technician A nor technician B
8. A scratch in a flexible bumper can be repaired with rigid plastic filler.
   a. True
   b. False
9. When painting a flexible bumper, technician A paints the difficult-to-reach areas first. Technician B paints the difficult-to-reach areas last. Who is right?
   a. technician A
   b. technician B
   c. both technician A and technician B
   d. neither technician A nor technician B
10. Primed flexible plastic parts should be sanded with _____ -grit sandpaper prior to sealing.

Chapter 15

Detailing

Objectives

After reading this chapter, you will be able to:
- Explain how to remove defects by color sanding and buffing.
- Explain how to apply pin stripes and decals.
- Explain how to install mouldings.
- Explain how to clean up a vehicle for delivery.

Key Term List

DOI

15.1 COLOR SANDING

Color sanding means to use a fine-grit sandpaper to remove surface defects before buffing. Color sanding can be wet or dry.

Wet color sanding can begin with 600-grit paper to remove serious defects such as runs. Progressively finer grits are used to eliminate the deep scratches and make buffing easier (Figure 15–1). A typical series of papers would be 1,000, 1,500 and 2,000 followed by buffing. If the defects are minor, only 1,500-grit paper may be needed. Use plenty of water. Squeegee off the sanding sludge to check on the surface.

FIGURE 15–1 *Wet-sanding clear coat before buffing.*

Dry color sanding also uses fine-grit paper. In this case, 1,200- or 1,500-grit paper is used on a DA (Figure 15–2). This method of color sanding leaves circular marks in the clear which are less noticeable than the long scratches left from wet sanding.

FIGURE 15–2 *Dry-sanding clear coat before buffing.*

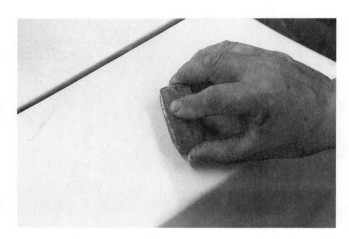

FIGURE 15–4 *Abrasive block in use.*

FIGURE 15–3 *This holder allows only a small area to be sanded.*

FIGURE 15–5 *Scraping clear coat runs with a razor blade.*

15.2 REFINISH DEFECTS

Refinish defects that may be removed by detailing are dirt on clear coat, runs in clear coat, and orange peel.

DIRT

Dirt in clear coat may be removed by sanding with fine-grit sandpaper. Figure 15–3 shows a special holder for 1,500-grit wet paper. This allows only a small area to be sanded. The technician can concentrate the sanding in only the area needed. If the dirt affects a large area, the entire panel can be sanded with 1,200-grit wet sandpaper and a sanding pad.

This can be followed with 1,500-grit paper. Sand until the dirt speck is flat and can no longer be felt.

Another method of removing dirt from clear coat is to use an abrasive block (Figure 15–4). Blocks are available in various grits; the grit is indicated by the color stamped on the end. The blocks work best when they are flat. To flatten the block, put a sheet of sandpaper with the same grit as the block on a flat surface, grit side up. Use water to lubricate as you move the block back and forth across the sandpaper to make a flat edge. The abrasive block can now be used on a dirt speck or other surface defect. It works like sandpaper. Lubricate the surface with water.

RUNS

Runs in clear coat are easy to remove with a razor blade. Be sure that the clear coat is thoroughly cured. Hold the razor blade as shown in Figure 15–5. Scrape

FIGURE 15–6 *Sandpaper wrapped around a squeegee allows only the run to be sanded.*

FIGURE 15–7 *Electric and air buffers.*

the excess clear coat off with the blade. Do not try to slice the run off, just scrape. Be careful not to dig into the surrounding clear. Scrape until the run is the same height as the surrounding clear. Sand all of the clear with 1,500-grit paper and buff.

To sand out a run, the biggest problem is avoiding the surrounding clear while sanding on the run. A way to avoid this problem is to wrap the sandpaper around a squeegee and use the edge on the run (Figure 15–6). This keeps the paper off the surrounding clear.

ORANGE PEEL

Orange peel can be removed by wet or dry sanding. If the orange peel is heavy, begin with 1,200-grit paper followed by 1,500-grit paper. Light orange peel may only need a 2,000-grit sanding.

15.3 BUFFING FRESH PAINT

CURE TIME

Check to see what the cure time is before buffing. In some cases, this can be as short as four hours air dry or as long as overnight. With urethane clear, it is best to buff as soon as it is allowed. If the urethane clear dries too long (as little as thirty-six hours in some cases), the surface will be too hard to sand and buff without difficulty.

BUFFERS

Buffers can be electric or air powered (Figure 15–7). The electric buffers produce more power and may be variable speed. Air buffers are lighter.

Electric Buffers

When using an electric buffer, always keep the cord away from the rotating pad. The simplest way to control the cord is to have the technician operating the buffer drape the cord over his shoulder. That way the cord will not get wrapped up in the rotating pad.

An electric buffer may rotate only at one speed. Often these types of buffers are operated by a switch. The switch is pushed to "on" and the pad spins. Other buffers are controlled by a trigger. The pad spins only when the trigger is pulled. Some buffers rotate at a high speed (2,500 rpm) and a low speed (1,500 rpm). The speed is determined by how far the trigger is pulled. A variable-speed buffer has a dial to set pad rotating speed. The range of pad rotating speed may be 2,500 to 800 rpm. On a variable-speed buffer, when the trigger is pulled, the pad rotates only at the dial set speed. Low pad rotation speed (1,500–1,200 rpm) works best when buffing urethane clear coat.

The weight of an electric buffer may lead to operator fatigue. However, the weight of the buffer may help in maintaining pressure when buffing heavy orange peel. To minimize fatigue when buffing an entire vehicle, switch positions often. For example, when buffing a hood, there is strain on the operator's lower back. Buff half of the hood, then switch to buffing a fender. If the operator squats, with his back straight, while buffing the fender, his back muscles will have a chance to recover.

Air Buffer

An air buffer air hose is heavy enough that it usually stays out of the way of the rotating pad. Still the operator should be mindful of where the hose is at.

The rotating speed of an air buffer pad can be regulated somewhat by changing the supply line air pressure. However at low rotation speed, the air buffer may be difficult to control.

The light weight of an air buffer forces the operator to bear down on the buffer when attempting to remove 1,200-grit scratches.

COMPOUNDS

Buffing compound lubricates the paint surface and dissipates the heat produced by the friction between the rotating pad and the paint surface. All buffing compounds contain abrasive. The coarser the abrasive, the more aggressive the cut. For example, if a clear coat is badly contaminated, it may be color sanded with 1,000-grit paper then 1,200-grit paper. A coarse buffing compound can be used to remove the 1,200-grit scratches; however, the coarse compound leaves scratches of its own in the clear coat. A new pad and a finer grit of compound may be used to remove the scratches left by the coarse compound.

Paint manufacturers recommend what type of compound should be used to buff. Always check the recommendations. Start with the least aggressive compound suggested. If this does not remove the problem, a coarser compound may be used. Be careful; if a coarse compound is used it may take hours of buffing with the compound to get the shine back in clear coat.

There should always be liquid compound between the pad and the paint. If the compound dries out, the lubrication is lost and the heat from the spinning pad may cut circular scratches—called swirl marks—into the paint. In extreme heat buildup, the paint may melt—called a paint burn.

PADS

Buffing pads can be made out of wood or foam (Figure 15–8). The wool can be natural or synthetic, coarse or fine. Wool pads may be single or double sided. When compared to foam pads, wool pads cut faster, requiring skilled operation to obtain a good finish. When buffing 1,200-grit color sand scratches, a coarse wool pad may be used with the appropriate compound. The operator may then switch to a fine wool pad and fine compound to finish.

FIGURE 15–8 *Wool and foam pads.*

Foam pads may be made out of reticulated or unreticulated foam. A nonreticulated foam pad runs hotter than a reticulated foam pad. The reticulated foam pad allows the friction heat to escape during use. Excessive heat can cause big problems, such as swirl marks or paint burns. Foam pads do not cut as fast as wool pads. If the paint surface has only a few defects, the slower cut is an advantage because there is less of a chance of rubbing through the clear coat. Foam pads may have the following designs:

- **Flat**. Produces the most heat, but difficult to control when held flat. Should be tilted to maintain control.
- **Waffle**. Air spaces allow this pad to run cooler than a flat foam pad. May be held flat, buffing a greater surface area than a tilted flat pad.
- **Concave**. This type of pad has a hollowed-out center. This tends to minimize heat buildup and keep the compound on the vehicle.

In summary, wool pads cut faster and require greater skill to avoid defects. Foam pads cut slower and are more forgiving.

Pads may be attached to the buffer by a threaded stud or a velcro backing plate. To change a pad with a threaded stud, disconnect the buffer from the power supply. If the buffer has a button to lock the pad shaft in place, hold the button down. If the button does not have a shaft lock, hold the shaft in place with a wrench. The pad can be removed if it is spun in the opposite direction from rotation. The operator does this by striking the edge of the pad with the palm of his hand. The old pad is threaded off and a new pad threaded on and tightened. The velcro pads are

FIGURE 15–9 Cleaning a wool pad with a spur.

FIGURE 15–10 Buffing the clear coat on a repainted pickup truck door.

changed by simply pulling off the old pad and centering a new pad.

Wool pads should be cleaned whenever dried compound builds up in the pad. Using a pad with dried compound will cause swirl marks. A spur is used to remove dried compound. The spur is held tight against the pad as the pad spins (Figure 15–9). Wool pads may also be cleaned in a washing machine. As a wool pad is used, the wool will gradually be removed. Discard any wool pads that have lost half of their original thickness due to use.

Foam pads may be cleaned and shaped with a wire brush. The brush is used in the same way as a spur. Foam pads may also be washed if they are contaminated. Discard a foam pad if more than one-eigth of the surface is damaged by tears.

If the surface has a few defects, it is much faster to remove them by color sanding before buffing. Once the clear coat has been color sanded, it is much easier to buff. Remember these rules when buffing:

- The paint surface should be clean. Grit or dirt could be buffed into the paint.
- Always keep the compound between the pad and the paint to reduce heat.
- Never buff on a body line or edge. Avoid antennas, bezels, stripes, and mouldings. If the buffer grabs them, it will break them. Pressure may be applied to the buffer when the compound is wet. As the compound dries during buffing, apply less pressure. Keep the pad clean. Dirt on the pad will make scratches in the paint. Always keep the buffer moving.

To buff fresh color-sanded clear coat, simply apply the compound to one panel surface. Spread out the compound with the buffer and buff the panel (Figure

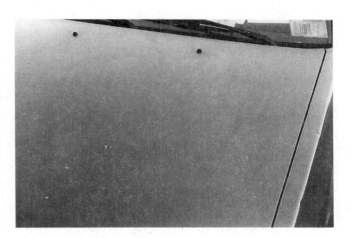

FIGURE 15–11 The paint on the hood of this red car is faded. It has no gloss or shine.

15–10). When all of the sanding scratches are removed and the surface has the desired distinctness of image (**DOI**), the buffing is done. A buffed fresh urethane surface will lack the intensity of an unbuffed surface. Do not excessively buff the clear coat.

Faded paint can be buffed to remove oxidized paint (Figure 15–11). Check the paint thickness with a mil gauge. If the paint thickness is less than 2.5–3.0 mils, there is not enough paint to buff. A repaint is the only way to repair this paint. The buffing procedure is the same as for fresh paint; color sanding will save time. When buffing faded paint, be especially careful of rub-through. There is only so much paint on the car. Each time the car is buffed, a little more paint is removed. Buff only enough to restore the gloss. See Table 15–1 for a summary of buffing procedures.

TABLE 15–1:   BUFFING PROCEDURE FOR PAINTS

Paint Type	Paint Condition	Procedure			
		Wet Sanding	Compounding	Machine Glazing	Hand Glazing
Refinish paints cured enamels/ urethanes* (air-dried more than 48 hours or baked)	1. Minor dust nibs of mismatched orange peel (light sanding)	1. Fine 1,500		1. Finishing material	1. Hand glaze
	2. Heavy orange peel dust nibs, paint runs or sags	2. Fine 1,200	2. Microfinishing compound	2. Finishing material	2. Hand glaze
Refinish paints: Fresh enamels/ urethanes* (air-dried 24 to 48 hours)	1. Minor dust nibs or mismatched orange peel (light sanding)	1. Fine 1,500	1. Microfinishing compound	1. Microfinishing glaze	1. Hand glaze
	2. Heavy orange peel, dust nibs, paint runs or sags	2. Fine 1,200	2. Microfinishing compound	2. Microfinishing glaze	2. Hand glaze
All factory applied (OEM)	1. New car prep or fine wheel marks				1. Hand glaze liquid polish
	2. Coarse swirl marks, chemical spotting, or light oxidation			2. Finishing material	2. Hand glaze
	3. Overspray or medium oxidation		3. Microfinishing compound (medium cut)	3. Finishing material	3. Hand glaze
	4. Heavy oxidation or minor acid rain pitting		4. Rubbing compound (heavy cut)	4. Finishing material	4. Hand glaze
	5. Dust nibs, minor scratches, or major acid rain pitting	5. Fine 1,500		5. Finishing material	5. Hand glaze
	6. Orange peel, paint runs, or sags	6. Fine 1,200 or 1,500	6. Microfinishing compound (medium cut)	6. Finishing material	6. Hand glaze

*Enamels/urethanes—as referred to in this chart—are catalyzed paint systems (including acrylic enamel, urethane, acrylic urethanes enamels, polyurethane enamels, and polyurethane acrylic enamels) and nonisocyanate-activated paint systems used in color or clear coats.

FIGURE 15–12 Hand-glazing the bedside of a pickup truck.

FIGURE 15–14 This Mustang has been painted with a tricoat white pearl. The clear coat was buffed and glazed.

FIGURE 15–13 Machine-glazing with an air buffer and foam pad.

15.5 — STRIPES

Most stripes are tape; a few are painted. Tape stripes may be obtained from OEMs or aftermarket. Often OEMs will sell only the vehicle stripes as a complete kit (all the stripes for a vehicle). Because most collision repairs involve only part of the vehicle, much of the stripe kit is not needed. Some OEMs sell stripes by individual panel or side of the vehicle. Aftermarket stripes are available in various widths, single, double and triple colors. If a needed stripe width is not available, a larger stripe can be slit to make the proper width. Usually collision repair shops put on tape stripes. Tape stripes are easy to apply. Follow these guidelines:

1. Clean off the surface where the strip will be placed. Remove wax, grease, compound, or grit.
2. Measure out the amount of stripe. If the entire side will be striped, do all of the panels on the side at once.
3. Remove the paper backing from the stripe. Leave the clear top on the stripe.
4. Start at one end, place the stripe, and pull the opposite end to tighten and form a straight stripe (Figure 15–15).
5. Lightly set the stripe.
6. Move away from the vehicle to see if the stripe is straight.
7. If the stripe is not straight, remove and reset.
8. When the stripe is straight, press it firmly to set.
9. Remove the plastic cover on the stripe.
10. Cut the stripe at the door openings.

15.4 — GLAZING

Glazing is used to remove the swirl marks caused by buffing. Glazing can be done by hand (Figure 15–12) or with a buffer and foam pad (Figure 15–13). When hand glazing, do not apply excessive compound. A ribbon about one inch long is enough to glaze a fender. Work the glaze onto the paint with a soft cloth. Remove the glaze before it dries. Machine glazing is similar. Do not apply too much compound. Work the compound in with the foam pad. A color-sanded, buffed, and glazed urethane is an outstanding finish (Figure 15–14).

FIGURE 15–15 Pinstripe installation on a repainted vehicle. Set the stripe at one end and pull tight to keep the stripe straight.

FIGURE 15–16 Masking tape has been installed as a straight reference line so a nameplate can be placed. Note that the masking was left in place while buffing. The tape on the quarter moulding shows a blend.

15.6 — MOULDINGS

New glued-on mouldings can be easily installed by first cleaning the surface to remove wax and grit. Next, check the opposite side of the vehicle to see where the moulding should be placed. On door mouldings, be careful not to set the front edge of the moulding too far forward. If the edge is too far forward, the moulding will hit the fender as the door is opened. Use masking tape as a reference line, either above or below the moulding location (Figure 15–16). Check the straightness of the reference line from a distance. Remove the protective tape from the mouldings. Set one end in place. Pull the moulding to keep it straight. Use the tape as a reference. Apply hard pressure to set the moulding in place. Remove the reference tape.

A reused glued-on moulding must have all of the tape removed from it. If the moulding is metal-backed and the metal was bent during removal, the metal backing should be ground off. With the metal backing removed, the moulding is pliable. Be sure all of the old tape is removed. Lightly sand the backside of the moulding with 220-grit paper. Wipe with a mild cleaner. Apply new double-faced moulding tape (Figure 15–17). Install the moulding in the same way as a new one.

Nameplates can be reused. In this case, the old adhesive needs to be scraped off. Lightly sand and apply double-faced tape (Figure 15–18). Cut to fit. An alternative method is to use moulding glue.

FIGURE 15–17 The technician is removing tape from a moulding so that fresh moulding tape can be installed.

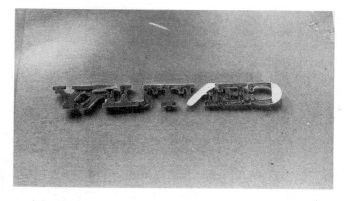

FIGURE 15–18 The adhesive has been scraped off this nameplate. Moulding tape was cut to fit. The white areas are areas where the backing has not been removed.

FIGURE 15–19   Wood grain decal on an S-10 Blazer.

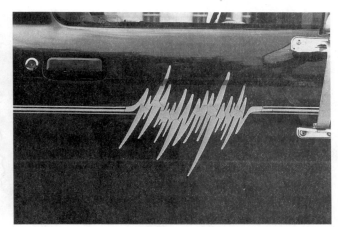

A

GRAPHICS

Graphics or decals are installed after all of the painting and buffing operations are completed. These overlays can be a wood decal (Figure 15–19) or intricate graphics (Figure 15–20) on a pickup truck. In any case, the procedure to install them is the same.

REPLACEMENT OF WOODGRAIN TRANSFER

Whenever woodgrain overlays are badly damaged, the only solution is replacement. They are available in suitable form from parts depots or service warehouses. When listing a woodgrain overlay replacement, be sure to have the correct body style, model year, car model, and panel name to which it is applied.

If the woodgrain transfer (decal) is to be installed on a new sheet metal panel or on a panel that required extensive straightening, remember that refinish operations must be completed before the transfer is applied. These transfers are never applied directly over bare metal or primed metal. However, new transfers can be applied over old transfers if the damaged areas are small, featheredged, and filled in with surfacer or spot putty to bring the surface up to the surrounding level. (Some woodgrain transfers are translucent, however, and different background colors will show through.) It is often necessary to remove exterior mouldings, handles, and lock cylinders before the transfer can be removed or installed.

To remove the woodgrain decal, use a heat gun to soften the adhesive on the transfer. Start at one edge and slowly peel the decal back. Keep the heat working over the area until the sheet is completely off.

B

FIGURE 15–20   Pickup truck graphics.

After the old decal is removed, repair the damaged metal and prime the repair. With either a new panel or a repaired one, sand the surface smooth and then clean with a wax and grease remover.

The first step in the reinstallation of the woodgrain decal is to make a template of the area to be covered. Using a sheet of masking paper, align it with the centerline of the moulding, attaching clip holes across the top of the panel. Tack-tape the paper in place.

With the template paper securely taped to the panel, mark the centerline of the panel on both the panel and the template. Smooth the paper flush against the panel and mark the front, rear, and bottom edges of the panel. If the woodgrain transfer on adjacent panels has a plank design, mark the top horizontal plank line on the front and rear edges of the panel.

Remove the template from the panel and lay it out on a flat, clean work surface. Measure three-quarters

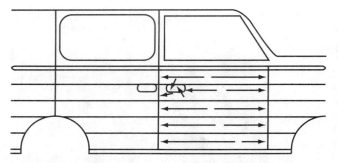

FIGURE 15–21 *Wood grain decal replacement on a minivan's right front door. Start at the top and work toward the bottom.*

of an inch out from the panel outline and mark another perimeter line. Oversizing the template this way will allow room for fitting the transfer to the panel. With a pair of scissors, cut out the template along the perimeter line. Mark the front edge of the template on the backside of the paper.

Now, roll out a sufficient amount of overlay and cut it to length. Lay the transfer face down on the work surface. Turn the template over and place it face down on the transfer. Make sure that the woodgrain is running right to left and that the horizontal planking lines on the template align with plank lines on the transfer (Figure 15–21). Trace the outline of the template on the transfer backing paper and cut the template to shape with a pair of scissors. Align the wood grain as closely as possible.

Hold the transfer cutout against the panel again. Carefully position the top edge of the transfer with the centerline of the trim clip holes and mark the centerline of the transfer with the centerline of the panel.

Lay the transfer face down again on the work surface and peel off the adhesive backing paper. With a sponge and a solution of water and liquid detergent, wet the adhesive side of the woodgrain overlay and the panel.

Align the decal with the clip holes and panel centerline. Lightly press the top of the transfer to the panel, making sure to align any plank lines or grain. With the transfer aligned, squeegee the center three or four inches of the transfer. Use an upward motion with the squeegee, forcing the liquid solution out along the top. This anchors the transfer into position.

Raise one side of the decal, and with short strokes, gradually squeegee the top edge of the transfer into place. Make sure the transfer edge stays even with the centerline of the moulding clip holes. Then squeegee the top edge with a long horizontal stroke. Repeat this procedure for the other side's top edge.

Then raise the transfer and, with the squeegee, press down another two or three inches in the center. Use overlapping horizontal strokes to bond another band of decal across the top of the panel. Progressively work down and across the top of the panel in this manner. If the decal gets tacky and sticks to the panel before it is pressed into place, break the grab with a fast, firm pull. Periodically, rewet the panel to decrease the tack as well as to make the transfer easier to position.

When reaching the edges of the decal, cut ninety-degree notches in the corners and V-shaped notches along the edges where necessary to fit the transfer to the panel. Avoid excessive pulling and stretching; the decal can tear.

Apply vinyl trim adhesive to door hem edges. Apply the adhesive sparingly to avoid a lumpy buildup under the transfer. Heat the edges of the door hem flanges and the transfer. Then wrap the transfer around the flange edge and firmly press it to the backside. Apply heat to any depressions or hole edges, and firmly press the transfer to ensure a good bond. Cut the excess decal away from panel edges and holes with a razor blade.

Inspect the application from an angle where light reflections will expose any irregularities. Pierce bubbles with a fine needle from an acute angle and press these down firmly. Reinstall all mouldings and other hardware.

SHOP TALK

Repaired vehicles returned to the collision repair facility because of a problem are known as comebacks. The appearance, durability, and function of the repaired vehicle are the collision repair facility's best advertisement. At the very least, a comeback annoys the customer. At worst, a comeback could destroy a facility's reputation. Always thoroughly inspect all vehicles prior to delivery. Do not allow the possibility of a comeback.

15.8 CLEANUP

After all of the repairs are finished, the car must be cleaned up before delivery. Start on the inside:

1. Clean the door jambs of compound and dust.
2. Blow out filler dust.
3. Wipe off all dust.
4. Vacuum the floor.

On the outside:

1. Clean the hood and trunk openings of compound and dust.
2. Wash the outside
3. Remove overspray
   a. from windows with a razor blade or window cleaner and a clay bar.
   b. from paint with a rubbing compound or water and a clay bar.
   c. from tires with lacquer thinner.
   d. from the wheelhousing by spraying with a black undercoat.

REVIEW QUESTIONS

1. Technician A color-sands with 600-grit sandpaper. Technician B color-sands with 1,200-grit sandpaper. Who is right?
   a. technician A
   b. technician B
   c. both technician A and technician B
   d. neither technician A nor technician B
2. Color-sanding can be done wet or dry.
   a. True
   b. False
3. Technician A says that dry color sanding leaves smaller scratches than wet color sanding. Technician B says that wet color sanding leaves smaller scratches than dry color sanding. Who is right?
   a. technician A
   b. technician B
   c. both technician A and technician B
   d. neither technician A nor technician B
4. Clear coat runs are easily removed by razor blade scraping.
   a. True
   b. False
5. Technician A says that foam pads run hotter than wool pads. Technician B says that wool pads run cooler than foam pads. Who is right?
   a. technician A
   b. technician B
   c. both technician A and technician B
   d. neither technician A nor technician B
6. What should be avoided when buffing a body line?
   a. body line
   b. edge
   c. antenna
   d. all of the above
7. Technician A uses glazing to remove swirl marks left from buffing. Technician B says that glazing can be done by hand or by machine. Who is right?
   a. technician A
   b. technician B
   c. both technician A and technician B
   d. neither technician A nor technician B
8. The first step in applying stripes is to clean the surface.
   a. True
   b. False
9. Technician A says that masking tape should be used as a guide before applying a glued-on moulding. Technician B says that all old tape must be removed from a moulding before it can be reused. Who is right?
   a. technician A
   b. technician B
   c. both technician A and technician B
   d. neither technician A nor technician B
10. Decals are applied _____ all painting and buffing.

GLOSSARY

Abrasive A substance used to wear away a surface by friction.

Acid rain A type of air pollution in which the rain has an acid pH.

Adhesion The ability of one substance to stick to another.

Aggressive cleaner The solvent used to remove contaminant from a surface before sanding.

Air drying Allowing paint to dry at ambient temperatures without the aid of an external heat source.

Air pressure Measurement in pounds per square inch of air compression.

Air transformer A device used to regulate air pressure.

Alternate A color that is slightly different from standard.

ASE The National Institute for Automotive Service Excellence.

Atomize The breakup of paint and solvent into fine particles by a spray gun.

Back blend The painting method in which each successive basecoat is stepped in two or more inches from the previous coat.

Back tape The masking method in which the edge of the masking tape is rolled to make a soft paint edge.

Basecoat The color coat in basecoat/clear coat systems.

Binder The component in paint that holds the paint particles together.

Blending The gradual shading of one color into another, used to hide a slight mismatch.

Block-sand To flatten a surface by sanding.

Body filler A heavy-bodied plastic material that cures very hard and is used to fill dents in metal.

Bullseye A depression or ring visible in the topcoat.

Catalyst A chemical that causes or speeds up a reaction when mixed with another substance.

Clear coat A topcoat on paint that is transparent so that the color coat underneath is visible.

CO Carbon monoxide.

Compressor A device used to compress air.

Conversion coating A special metal conditioner used on bare metal to prevent rust.

Corrosion The chemical reaction of oxygen, moisture, and steel that results in rust.

Crossdraft Air movement in a spray booth, in from one end of the booth and out through the other end.

Cross link The curing process in catalyzed paint.

Cure The process of drying or hardening of a paint film.

DOI Distinctness of Image.

Drying The process of change in a coat of paint from the liquid to the solid state due to the evaporation of the solvent, a chemical reaction of the binding medium, or a combination of these.

Enamel A type of paint that dries in two stages: first by the evaporation of the solvent and then by the oxidation of the paint film.

EPA Environmental Protection Agency.

Epoxy primer The undercoat that binds to bare metal.

Featheredge Tapering of broken paint edges with sandpaper.

Filler Any refinishing material used to level a surface.

Flash off The time required for a solvent to evaporate.

Flash time The waiting time, in minutes, between coats of paint.

Flex agent An elastomeric additive that allows paint to remain pliable after drying.

Force-drying Heating a freshly painted surface to speed drying.

Front blend The painting method in which each successive basecoat is stepped out two or more inches from the previous coat.

Galvanized Metal coated with zinc.

Gravity feed Paint feed method in which the paint flows to the fluid tip because the paint cup is above the fluid tip.

Grit A measure of the size of particles on sandpaper or discs.

Guide coat Dusting of paint to aid in block sanding primer.

Hardener A curing agent used in plastics.

Hazardous waste Unusable by-product that could harm the environment if not properly disposed of.

High crown A panel that is curved.

HVLP High volume, low pressure.

Hydrocarbon (HC) A molecule composed of hydrogen atoms and carbon atoms.

Isocyanate This is the principal ingredient in ure-thane hardeners. Because this ingredient has toxic effects on the painter, the painter is always advised to wear a respirator approved by NIOSH.

Lacquer A type of paint that dries by solvent evapo-ration. Can be rubbed to improve appearance.
Low crown A panel that is relatively flat.

Manometer Device used to measure air pressure differences.
Mass tone The dominant color in a tint.
Melt Paint method in which clear coat overspray is dissolved to make an invisible edge.
Metal conditioner A chemical cleaner that removes rust and corrosion from bare metal and helps prevent further rusting.
Metallics Finish paint colors that contain metallic flakes in addition to pigment.
Mil A measure of paint film thickness equal to one one-thousandth of an inch.
Mild cleaner Solvent used to remove contaminant from a surface before painting.
MSDS Material Safety Data Sheets.

OEM Original Equipment Manufacturer.
Orange peel A paint surface that is bumpy and rough.
Organic A carbon-containing compound.
OSHA Occupational Safety and Health Administra-tion.
Overlap The amount of the spray pattern that cov-ers the previous spray stroke.
Overspray Paint that falls on the area next to the one being painted.
Oxidation The combining of oxygen from the air with a paint film. One principal cause of alkyd and acrylic enamel drying.

pH The measurement of acidity or alkalinity of a so-lution: 0 to 6.9 is acid, 7.1 to 14 is base, and 7.0 is neutral.
Pigment Material in the form of fine powders used to impart color, opacity, and other effects to paint.
Polychrome Metallic paint.
Pot life The amount of time, measured in hours, within which paint must be sprayed after hardener is added.
Pressure feed A paint feed method in which paint is pushed to the fluid tip because the paint cup is under pressure.
Primer The undercoat that is applied to bare metal to promote adhesion of the topcoat.
Putty A material used to fill small holes or sand scratches.

Reducer The solvent used to dilute enamel.
Refinish To repaint.
Respirator A device worn over the mouth and nose to filter particles and fumes.
Rustproofing Methods or materials used to prevent the corrosion of steel.

Sealer Undercoat used to promote adhesion and provide uniform color.
Seam sealer Materials used to fill gaps between panels.
Self-etching primer Undercoat that treats and bonds to bare metal.
Semi-downdraft Air movement in a spray booth in from the ceiling at one end and out at the opposite wall.
Shelf life The amount of time during which hardener can be used once opened.
Single-stage compressor Device that compresses air to final pressure in one stroke.
Siphon feed Paint feed method in which paint is pulled from the cup by suction.
Solvent A liquid capable of dissolving another substance.
Spectrophotometer Device used to analyze color.
Spray booth Enclosed area to contain and remove spray paint vapors.
Spray gun Device that atomizes liquid paint.
Standard A reference color, used for comparison.
Surfacer High-solids undercoat designed to fill scratches.

Thermoplastic Type of plastic that can be reshaped with heat.
Thermoset Type of plastic that cannot be reshaped with heat.
Thinner The solvent used to dilute lacquer.
Tint Colors combined together to make paint.
Topcoat The last or final color or clear coat.
Toxicity Pertaining to poisonous effect.
Translucent A substance that allows some light to pass through, partially transparent.
Transparent A substance that allows light to pass through.
Two-stage compressor Device that compresses air to an intermediate pressure in one cylinder, then to a final pressure in another cylinder.

Undercoat A first coat, primer, surfacer, or sealer.
Underhue A secondary color in a tint.

Vaporization The conversion of solvents into gases during spray painting.
Viscosity The consistency or body of a paint.
VOC Volatile Organic Compound.
Volatile Capable of evaporating easily.

Index